# MODELING HUMAN ERROR IN STRUCTURAL DESIGN AND CONSTRUCTION

Proceedings of a Workshop sponsored by
the National Science Foundation

Approved for publication by the
Construction Division of the
American Society of Civil Engineers

Ann Arbor, Michigan
June 4-6, 1986

**Edited by Andrzej S. Nowak**

ASCE 1852

Published by the
American Society of Civil Engineers
345 East 47th Street
New York, New York 10017-2398

**ABSTRACT**

Human error is recognized as the major cause of structural failure. Moreover, it may greatly exacerbate the damaging effect of other hazards, such as earthquakes. The objectives of this workshop were to take status of research results up to date in order to identify the most effective approach for the near future; to identify, in particular, the promise and difficulties of the scenario approach; and to determine the degree of background preparation necessary for effective improvement of structural reliability in the long range, i.e. optimal control of the process of structural production. Top international experts were brought together with the aim that a coherent strategy for research and professional activity for the next five to ten years can be identified and agreed upon by the most active workers on the subject of human error in structures. The technical papers contributed by the participants deal with various aspects of the problem. The discussion and findings of three working groups (occurrence of errors, reliability models and error control strategies) are summarized in the reports which are included in this book.

This material is based upon work supported by the National Science Foundation under Grant No. MSM-8606218.

Any opinions, findings, and conclusions or recommendations expressed in this publication are those of the writer(s) and do not necessarily reflect the views of the National Science Foundation.

**Library of Congress Cataloging-in-Publication Data**

Modeling human error in structural design and
construction.

Includes bibliographies and indexes.
1. Structural stability—Congresses 2. Safety factor in engineering—Congresses. 3. Reliability (Engineering)—Congresses. I. Nowak, Andrzej S. II. National Science Foundation (U.S.) III. Title: Human error in structural design and construction.
TA656.M63 1986 624.1'771 86-22144
ISBN 0-87262-558-3


Library of Congress Catalog Card No: 86-22144
ISBN 0-87262-558-3
Manufactured in the United States of America.

# CONTENTS

# ACKNOWLEDGMENTS

The Committee acknowledges with deep appreciation the work of those who made the presentations. It was a major effort on their part.

The financial support of the National Science Foundation, with Gifford Albright, Program Director, is gratefully acknowledged.

Thanks are due to J. Czernecki, J. Melgarejo, J. Kayser, A. Arafah, and S-C. Ting for their help in editing and to G. Singleton, R. Mantel, S. Glinke, and K. Helm for typing the texts.

All of the participants contributed generously of their time. The extent of enthusiastic involvement and the full attendance at all the sessions is a testimony to their contribution and the importance of the topic.

It is hoped, now, that the professors and other researchers will submit their proposals, that industry will join in supporting the research and provide advice, and that government agencies (and especially the National Science Foundation) will have sufficient financial resources to support the massive undertaking needed to efficiently control human errors in design and construction.

ORGANIZING COMMITTEE

Andrzej S. Nowak—Chairman
University of Michigan

Robert I. Carr
University of Michigan

Ted V. Galambos
Univeristy of Minnesota

Niels C. Lind
University of Waterloo

# OPENING REMARKS

Gifford H. Albright*

On behalf of the National Science Foundation (NSF) I would like to indicate how pleased we are to be associated with this important endeavor. I believe that the NSF and the civil engineering community can significantly increase structural reliability, through a better understanding and control of failure causes. Human error has been identfied as the major cause of structural failures.

The National Science Foundation, through its program on Structures and Building Systems supports research to enhance the development of knowledge basic to the design and construction of safe, long-lived, maintenance-free, economical structures and facilities, and research basic to the design, construction, and efficient operations of buildings. Emphasis is placed on innovative research and attention to the dissemination of research results. Research areas of interest are included within three general categories: structural systems, construction processes, and building systems. Structural reliability and human error are a major research interest in all of these.

Research in structural systems includes predictions of structural loadings from many sources including wind, methods of analysis of buildings and structural systems and their components under fixed, steady-state, and transient loads; active and passive control of structural responses to extreme loading conditions; structural reliability and optimizations, knowledge-based expert systems for structural design, and evaluation of existing structures and materials.

Research in construction processes includes innovative construction processes and systems, advances in construction materials, construction engineering risks and uncertainty, organization and management of human and material resources in the construction process, knowledge-based expert systems and robotics in construction.

Research in building systems includes integration and optimization of building systems—structural, material, building services, communications, and equipment; building evaluation and diagnostics; physical environment and integrated systems engineering to optimize security and safety including accidents and building failures; conservation of resources and building re-use, knowledge-based expert systems and computer-aided design of integrated building systems.

I believe that a large number of important research topics can result from this Workshop.

---

*Director of Structures and Building Systems, National Science Foundation, Washington, D.C. 20550.

## INTRODUCTION

Andrzej S. Nowak*, M. ASCE, and Niels C. Lind**

Human error is recognized as the major cause of structural failure. Moreover, human error in design or construction may greatly exacerbate the damaging effect of other hazards, such as earthquakes.

Structural reliability theory has advanced considerably in the last twenty years and has found many applications to structural design standards and procedures. Through the improved understanding of the risk of structural failure that it provides, it has become possible to achieve more optimal investment in that part of structural safety that is effectively controlled by the load and resistance factors.

The research front has therefore increasingly turned its attention towards human error in structures as an effective object for control. From sporadic efforts in the late 1970's (Pugsley 1973, Knoll 1976, Sibly and Walker 1977, Allen 1977, Rackwitz 1977, Matousek 1977, Brown 1979 and Nowak 1979) the breadth and the importance of the human error problem as a major issue in structural safety became recognized around 1980. Since then, several major studies of the problem have been underway in America, Europe and Australia. Results of these studies are now available as a basis for future work. Simultaneously, human error has also been recognized as a major issue in the safety of other man-made facilities and systems. The perception of risk has been drawn to the importance of human factors through incidents or disasters such as Three Mile Island, Bhopal or, more recently, Chernobyl. For effective control of the safety of any man-made system an understanding of human error is now essential. The concurrent changes in emphasis on product liability has increased the expectations of flawless performance of engineers' systems.

Research into human error in structures has approached the subject from two opposite main directions: fundamental studies and frameworks for application. Fundamental studies seek to deepen the understanding through disaggregated error statistics (Matousek 1977, Smith 1977), phenomenological models (Lind 1979) or heuristic models (Knoll 1976). They

--------------------

*Associate Professor of Civil Engineering, University of Michigan, Ann Arbor, MI 48109.

**Professor of Civil Engineering, University of Waterloo, Waterloo, Ontario N2L 3G1, Canada

range from experimental studies of the commission of errors (Melchers and Harrington 1982, 1984) to simplified mathematical models (Lind 1983) or even stochastic process models (Ditlevsen and Hasofer 1983).

Several studies of possible frameworks for application to structural engineering or construction have been developed simultaneously. In the pioneering study by Pugsley (1973) the notion of error-prone stuctures was introduced, and a framework was sketched to identify such structures by a combination of attributes. This work was refined by Blockley (1975, 1977) who has also explored the use of fuzzy set concepts in the subject area. Fox (1981) continued this work, identifying the most significant attributes and suggesting a point scheme for estimating the magnitude of the probability of failure due to human error. Lind (1983) gave a framework for determining the optimal allocation of engineering time to design modeling, materials testing, and inspection of construction in relation to a given target level of design reliability. In several studies Nowak and Carr (1985) have given a systematic framework that relates human errors to their effect on structural reliability via sensitivity coefficients.

On the level of professional societies there have also been several efforts to search for a framework for application. Thus, IABSE sponsored a workshop on Quality Assurance (Schneider, 1983) focusing largely on human error. Other workshops that have been effective in defining the problem and placing it on the research agenda have been sponsored by the National Research Council of Canada (NRC, 1979), and the Institute for Risk Research, University of Waterloo (Grigoriu, 1983).

Application of these various concepts and ideas to structural practice is, of course, still far from being accomplished. Much research remains to be done, but there is no clear consensus of the best directions and emphasis. Research has until the present proceeded from individuals' initiatives and some aspects that are clearly of fundamental importance remain unexplored, for example the sociology of construction organization. It has frequently been pointed out that good human relations and communications with clear definitions of responsibilities is important in error control, but this aspect has not been explored in any depth. It has also been suggested that the concept of scenario can be exploited to develop an effective error control formalization, but individual contributions to this subject are few and sporadic.

The problem, then, is to bring the top international experts together with the aim that a coherent strategy for research and professional activity for the next five to ten years can be identified and agreed upon by the most active workers on the subject of human error in structures.

The workshop has three objectives:

- to take status of research results up to date in order to identify the most effective approach for the near future;

- to identify, in particular, the promise and difficulties of the scenario approach;

- to determine the degree of background preparation necessary for effective improvement of structural reliability in the long range, i.e. optimal control of the process of structural production.

The first five years of this decade have seen the exploration in some depth of the gross error phenomenon and a beginning has been made towards a global systematic view of the production of structures. The objective is now to establish a rapport between these two complementary aspects of new structural engineering theory. The expected outcome is an identification of some promising approaches, giving a unified conceptual framework for future research.

Side by side with this quality assurance viewpoint of construction, it appears that the concept of scenarios may provide for the transition in design process to unify the conventional structural safety controlled by load and resistance factors, and safety against human error. Scenarios seem more appropriate to forecasting, economic and social planning, and public policy assessment, but it is widely believed that they hold promise for structural design by providing a unifying rational for design, construction, materials testing and inspection. Whether this promise can be realized depends on several factors. For example, what are the characteristics of a "good" scenario? Is such a scenario crucial to one particular mode of failure? How is a complete set of scenarios generated? How can sufficiency be verified? How are the quantities that appear in the scenario to be chosen (for example: What should be the return periods of the loads)? Finally, can the scenario approach offer anything more than a formalization of the present engineering rationale?

Structural engineering theory has during the past century been dominated by the needs of quantitative analysis: determine the loads and find the stresses, and special tools are now available to facilitate this task. This need has, in turn, dominated the formal education of the structural engineer, and researchers in structural engineering have mainly been prepared for work in structural mechanics. Control of human error in structures may require a broadening of the educational background of the researcher and ultimelately a change in emphasis if not a revolution in the formal education of structural engineers. It is a final objective of the workshop to identify the needs of the education of the structual engineers and researchers of the future, necessitated by developments in reliability theory, quality assurance, and acceptable performance. Also, it is the aim that a specific concept for implementation of these developments be laid out.

The Workshop was held on June 4-6, 1986 at the University of Michigan in Ann Arbor. About 30 active researchers and leading structural engineers were brought together for a three day period.

The first day of the Workshop was devoted to presentations and discussions of recent progress. The participants submitted their written contributions prior to the meeting. These contributions are included in the proceedings.

On the second day the participants were divided into three working groups, each developing specific steps toward the objectives.

The final day plenary session was devoted to critical review of the working group reports and planning of future work.

Each of the papers included in the Proceedings has been accepted for publication by the Proceedings Editor. All papers are eligible for discussion in the appropriate ASCE journals. All papers are eligible for ASCE awards.

References

Allen, D.E., "ACI Error Survey Canadian Data", Building Research Note, No. 123, National Research Council of Canada, Ottawa, July 1977.

Blockley, D.I., "Predicting the likelihood of Structural Accidents", Proc. Instn. Civ. Engrs., Part 2, 59, Dec. 1975, pp. 659-668.

Blockley, D.I.,"Analysis of Structural Failures", Proceedings, Instn. Civ. Engrs., Part 1, 62, February 1977, pp. 51-74.

Brown, C.B.,"A Fuzzy Safety Measures", ASCE, Engineering Mechanics, October 1979, pp. 855-872.

Ditlevsen, O. and Hasofer, A.M., "Design Decision Model Considering Mistake-Proneness", Report No. 273, The Technical University of Denmark, September 1983.

Fox, S.R.,"Predicting the Proneness of Building to Gross Errors in Design and Construction", A Thesis Presented to University of Waterloo, Department of Civil Engineering, March 1982.

Grigoriu, M.D., ed. "Risk, Structural Engineering and Human Error", Proceedings of a Symposium on Technology and Risk, University of Waterloo, Waterloo, July 1983.

Knoll, F., "Commentary on the Basic Philosphy and Recent Developments of Safety Margins", Canadian Journal of Civil Engineering, Vol. 3, No. 3, 1976, p. 409-416.

Lind, N.C., "Optimization, Cost-Benefit Analysis, Specifications", Proc. 3rd Int. Congr. Appl. Stat. and Prob. in Soil and Structural Engineering, v.3, Unisearch, Sydney, NSE, 1979, pp. 373-384.

Lind, N.C.,"Management of Gross Errors", Fourth International Conf. on Applications of Statistics and Probability in Soil Structural Engineering, Italy, 1983.

Lind, N.C., "Models of Human Error in Structural Reliability", Structural Safety, Vol. 1, No. 3, April 1983, pp. 167-175.

Matousek, M., "Outcome of a Survey on 800 Construction Failures", IABSE Colloquium on Inspection and Quality Control, July 1977, Cambridge, England.

Melchers, R.E. and Harrington, M.V., "Human Error in Simple Design Tasks", Research Report No. 3/1982, Monash University, Australia.

Melchers, R.E. and Harrington, M.V., "Human Error in Structural Reliability - Investigation of Typical Design Tasks", Research Report No. 2/1984, Monash University, Australia.

Nowak, A.S., "Effect of Human Errors on Structural Safety", Journal of the American Concrete Institute, September 1979, pp. 959-972.

Nowak, A.S. and Carr, R.I., "Classification of Human Errors", Proceeding of ASCE Symposium on Structural Safety Studies, Denver, May 1985, pp. 1-10.

Nowak, A.S. and Carr, R.I., "Sensitivity Analysis for Structural Errors", The Journal of Structral Engineering, ASCE, Vol. 111, No. 8, August 1985, pp. 1734-1746.

Pugsley, A.G., "The Prediction of the Proneness to Structural Accidents", Struct. Engr., 1973, 51, No. 6, June, pp. 195-196.

Rackwitz, R., "Note on the Treatment of Errors in Structural Reliability", Technische Universitat Munchen, Laboratorium fur den Konstrucktiven Ingenieurbau, Heft 21, 1977, pp. 23-35.

Sibly, P.G. and Walker, A.C., "Structural Accidents and their Causes", Proc. Instn. Civ. Engrs, Part I, 62, May 1977, pp. 191-208.

Smith, D.W., "Why Do Bridges Fail", Civil Engineering, American Society of Civil Engineers, November, 1977, pp. 58-62.

Schneider, J., ed., "Quality Assurance Within the Building Process," Proceeding, Vol. 47, IABSE Workshop, Rigi, Switzerland, June 1983.

# HUMAN ERROR AND STRUCTURAL PRACTICE

David E. Allen*

## Abstract

Following the ideas of De Bono on how the human mind works, paper discusses how mistakes are made, how mistakes are found, and applies these ideas to structural engineering practice and failure experience.

## Introduction

"To err is human, to forgive divine"

At the recent Rigi workshop on quality assurance (Schneider 1983) it was agreed that people are the most important elements of the building process and that human factors are the most important factors in preventing structural failure. This paper explores some ideas (De Bono 1971, 1976) on how the mind works, how it makes mistakes, how mistakes are found, and applies some of these ideas to failure experience. Can we do better by knowing more about ourselves?

## The Mechanism of Mind

Like anything else the mind can be modelled. A model is a simplified representation of a system in our minds that never quite fits reality but may be useful because it is something we understand and can manipulate and, in so doing, increase our understanding of the system. Science and engineering are full of such models.

The mind is a system that processes information. De Bono (1971) suggests mechanical models of the mind based on the concept of thinking as a mechanism of activation of a special memory-surface and of flow of activation across it. One of these models which appeals to me consists of jelly which has a flat surface at birth and onto which drops of hot water, representing information from the environment, are spooned in a pattern sequence. The water melts the jelly forming depressions on the surface. Water subsequently spooned onto or adjacent to a depressed area flows into it and deepens it; thus incoming information is affected by the information already assimilated and rein-

---

*Research Officer, Institute for Research in Construction, National Research Council, Ottawa, Ontario K1A OR6, Canada

forces certain parts of it. In this way a special memory-surface is built up. Even when no information is coming in there is a flow of activation (i.e. water) across the memory surface in accordance with the contours formed by the depressions; this flow in turn deepens the channels of flow and increases the rate of flow. This flow may pause briefly in a depression displaying a pattern in the conscious mind.

The sculpted surface of the jelly corresponds to long-term memory - things recalled if we jog our memory. Short-term memory - things we forget after a while, is represented on the jelly surface by wetness left over from recent drops of information from the environment which has not yet evaporated but which may be reactivated by a connected pattern of thought. Occasionally the wet spots in high areas can open up a new channel and pattern of thought, but it gets more difficult if the valleys are deep.

One property of the jelly model is that the activation or water covers a small finite area of the memory surface at any one time (concentration property); this applies to handling incoming information as well as thinking or activation flow. If, for example, a confusion is presented, the mind will only pick out patterns that it can deal with. On the other hand, if a word is presented, it will trigger a whole pattern on the memory surface. Another property of the model is that the attention area triggered determines, by its contours, the subsequent flow of activation (flow property); the sequence of incoming information therefore makes a difference to the pattern of thinking established. A third property of the model is that existing patterns get more firmly established through repetition and it becomes more difficult to establish new patterns from the incoming information (familiarity property); only when the gap between the pattern established on the memory surface and the environment is large will the pattern get changed. Finally activation flow links patterns together into unified patterns which lose some of the detail but gain in generality (abstraction property). It is this property of the human mind that makes it so powerful. Words, for example, are named abstract ideas and sentences are a connected flow of named abstract ideas.

The familiarity property of the mind (firmly established patterns) allows it to deal with situations quickly but also leads to thinking mistakes. The ability of the mind to trigger a pattern and flow from only a clue from the environment or a word leads to guessing mistakes. Despite these faults the human mind is very good at processing information (mainly by eliminating most information as useless and by simplifying and generalizing the rest into established patterns) but it is not good at storing it. The computer, on the other hand, is good at storing information but not good at processing it.

An important feature of the mind is emotional bias which affects thinking in response to internal needs such as hunger. This biasing factor tends to trigger patterns on the memory surface (such as food) and subsequent flow related to the internal need. I believe that motivating factors such as dissatisfaction, curiosity and fear of failure act as internal biasing factors in the same way as hunger and are therefore very important factors in the thinking process.

The mechanism of mind is summarized by the following quote from p. 121 from (De Bono 1971)

"The behavior of the special memory-surface is pretty effective. It can select and it can ignore. It can fragment into units or create new units out of fragments. It can extract patterns from confused material and it can abstract standard patterns from material that shows variation. It can represent complete patterns with a symbol and it can elaborate a symbol into a complete pattern. All this adds up to a remarkable information processing system. But the system also has its limitations. These limitations are inseparable from the advantages, since they are the advantages looked at in a different way. For instance the ability to establish firm patterns implies that changing a pattern may be very difficult. The ability to create new patterns means that these new patterns may be regarded as really existing in the environment when in fact they do not"

Practical Thinking

The basic biological purpose of thinking is to enable a living organism to survive by getting the things it needs and keeping clear of the things that are dangerous. Humans got very good at this compared to other creatures, so much so that we now elevate this activity to more creative interests such as music. For everyday practical concerns, however, it is basically a matter of knowing what to do about a situation that confronts us. There are three basic practical thinking processes for doing this -instinct (a prewired reaction to the environment), learning (either first hand or second hand, the result of which is do/don't rules) and understanding. For structural design and checking learning is important, but understanding is most important.

Understanding is a thinking process that changes a new and unfamiliar situation into one or more familiar ones which we can handle. It involves a process of going from one idea to another, or puzzling over it; such a pattern flow has already been described by the jelly model. Some of the advantages of the understanding process are: (i) it can be explained to others; (ii) it converts many new situations into a few familiar ones, i.e. it multiplies knowledge. In structural analysis, for example, the structure is broken down into component models simplified in the mind, each of which we can deal with using a few basic principles taught to us in school. Some of the disadvantages of understanding are: (i) one is limited to the ideas available on the memory surface; (ii) it is personal and subjective and, therefore, (iii) it distorts reality. These disadvantages are unavoidable and arise from the properties of the human mind described above. Understanding is still the best tool around, and often the only tool, for dealing with practical questions such as diagnosis of failure, structural design and checking.

Diagnosis is trying to understand a situation which does not work. Design, on the other hand, is creating a situation (from images on the memory surface) and trying to understand (checking) whether it will work. The same thinking process of understanding applies to both.

## Making Mistakes

Mistakes cannot be avoided because they arise directly from the way the mind works. A useful experiment to show up the errors in the thinking process is to try to do things in a great hurry. More mistakes occur because our actions are based on internal patterns triggered off by incomplete information - no time to puzzle over it and fully understand it. As we have more time to think about it we make fewer mistakes. Given enough time, however, we still make mistakes. This occurs either because we are distracted by something else or because the information is incomplete. The memory surface provides ideas to replace the missing information so as to arrive at an explanation which is needed to take action.

De Bono (1976) describes the following types of mistakes:

Monorail Mistake. This occurs when you follow along a single track from one idea to another ignoring the qualifying factors what make the idea work. An example is the importation of rabbits to Australia because they provide good sport hunting without consideration of their control.

Magnitude Mistake. Here one looks at the named idea but not the magnitude. Children often make magnitude mistakes, as do people confronted with a situation for which they have no experience.

Mis-fit Mistake. Something is recognized as familiar which in fact is not so; the idea is based on the perception available which does not fit reality. An example is catching up to a stranger from behind who you think you know. I made a mis-fit mistake when greeting the identical twin of an old friend of mine.

Mis-out Mistake. One considers only part of the situation and yet reaches conclusions that apply to the whole situation. What happens is that either attention is paid only to a part of the information or that only part of the information is available and one is not aware that there is additional information which changes the situation. Examples include missing a turnoff while driving because your mind is elsewhere or being unaware of egg on your face.

Must-be Mistake. A conclusion reached by available perception is fixed by arrogance. Part of the reason for this is that we are taught in school that 'being right' is good (pass) and so 'being wrong' is bad (fail). This is not really a type of mistake but an inability to find one and correct it.

To summarize, practical thinking involves coming to a conclusion before it can actually be checked out. Being right is a feeling of being right (not a fact) because this is what one acts upon. You make a mistake because you feel you are right; it is not deliberate.

## Finding Mistakes

The best way to recognize mistakes (and correct them) is to develop thinking tools for this purpose. In order to find mistakes one pays attention to the situation, looks for clues which generate ideas and gradually one comes to an understanding which tells you that there is a mistake.

Attention. A basic property of the mind is that one cannot pay attention to everything. What one does when confronted with a situation which is complex is to carve out areas of attention: (i) in space, paying attention to a part or a feature of the whole picture; (ii) in time, paying attention only to part of a sequence of events and (iii) in depth, paying attention only to some details.

Clues. Having carved out an attention area one then looks for clues. Clues are specific features of the attention area which generate possible ideas. Ideas are simple patterns or concepts on the memory surface which one can manipulate to come to an explanation. There are generally many clues which can generate many ideas and, hence, many explanations. This is not useful. Also some of the ideas do not make sense. What one does is shuttle back and forth between ideas in the mind and clues (and sometimes attention areas) in the environment in search of a set of ideas or a concept which provides an explanation that makes sense. Clues are therefore used not only to generate ideas but to confirm some and eliminate others. Diagnosis, for example, is based on this approach.

In practical situations one searches for a simple set of ideas or a concept that makes sense and explains the situation sufficiently well; inevitably, this will distort the real situation somewhat. Usually the distortion or mismatch does not affect the result since the explanation is near enough, but sometimes the mismatch is great and this can happen if significant clues are overlooked or if one comes to a single explanation too hastily.

Finding mistakes therefore involves shuttling back and forth between ideas and clues and attention areas. Previous experience with similar situations is very helpful because it provides ideas on the memory surface as well as experience with this type of thinking process. Imagination, the ability to access images, patterns or ideas on the memory surface and to move from one to another, also helps. Searching for 'hazard scenarios' is useful in this regard for design checking.

Often the human mind is not very forthcoming, however. In fact some of the best thinking occurs when you are not trying to think about the situation but about something else, for example while driving or jogging. De Bono describes thinking tools to help generate this type of 'lateral' thinking.

Communication also helps find mistakes because one makes use of more than just one memory surface, i.e. more than one database of experience; after all one can only generate the ideas that are available. Taking your design to the person next door often finds mis-out mistakes. Taking your ideas to experts outside your area helps avoid monorail mistakes.

One of the difficulties of communication is disagreement. This often occurs because the initial attention areas are different; recall that the choice of attention area determines the thinking (flow of activation) that follows. It would be very surprising if two people starting with different attention areas ended up with the same conclusion. Instead of denying the existence of one or other viewpoint, it is better to compare initial attention areas (starting points, assumptions) and see where they lead to. Connections can often be made and the two viewpoints improved into one.

## Failure Experience

A quick review of failure experience indicates that all types of mistakes occur. Moreover, usually a number of people are involved in each failure, each making a different type of mistake. Finally these mistakes are often found in time if the people unknowingly making them communicate with each other.

For example, the falsework supporting a bridge under construction collapsed because of a monorail mistake by the contractor (who did not understand a qualifying factor that made the falsework design work in a previous application but not in the one that failed) combined with a mis-out mistake by the bridge engineer (who did not pay full attention to the falsework design), engendered by a lack of communication between the two.

Monorail mistakes often occur when a change is not communicated to those on the project team who understand the qualifying factors that make the changed situation work. Changes often have ramifications beyond the direct solution of the problem that brought them about.

Mis-out mistakes due to lack of attention in space occur when only part of the structure is given attention (e.g. main members and not certain details); also when the structure or detail is looked at in one direction but not in another (three-dimensional problem). Sometimes no one steps back and looks at a whole project.

Mis-out mistakes due to lack of attention in time occur when attention is given to the design of a structure as a finished object but not to how it will be put together, how it will be inspected during use, or, how it will be disposed of at the end of its use. A lack of construction plan figures prominently in many construction collapses.

Magnitude mistakes sometimes occur during construction. In one case formwork removal was based on visual appearance and hardness of the concrete determined by stabbing it with a reinforcing bar. Inadequate construction procedures or unclear drawings generate such mistakes.

Even mis-fit mistakes occur. In one case engineers incorrectly diagnosed failure of flimsy tabs supporting a heavy ceiling to corrosion.

Clues are sometimes neglected. Sometimes the clues are obvious such as a partial failure which is disregarded. Sometimes concerns expressed by non-experts about the safety are not taken seriously. Failure to apply lessons from previous failures falls into this category.

Sometimes must-be mistakes occur because people are 'dazzled' by the someone's reputation to the point of uncritical acceptance of his advice.

Finally the use of simple concepts in design and construction procedures is important in preventing the generation of mistakes. The reason is that the human mind (not only the one that conceived it) can understand why it works.

References

De Bono, E., "The Mechanism of Mind", Pelican Books, 1971,

De Bono, E., "Practical Thinking", Pelican Books, 1976.

Schneider, J.,ed. IABSE Workshop, Rigi Report, June 1983.

# INCOMPLETE DESIGN PARADIGMS

Colin B. Brown, M. ASCE*

Abstract

Professionals establish informal paradigms for the conduct of design activities. These paradigms may not be suited to changing circumstances and become more mischievous than helpful. An example is the consideration of beam stiffness in the design of suspension bridges.

## Introduction

The activity of design of structures is not deductive and results in an absence of solution uniqueness. However, professional acceptance of paradigms for the design activity tends to replace the deductive process and to lead to acceptable solutions. The requirements of good paradigms are:

a) That they be implicitly or explicitly adhered to by the profession.

b) That their ambiguity and vagueness of form ensure the continuing use of the model.**

Codes are explicit requirements which tend to be followed by the profession but are seldom vague enough to ensure continual validity. The commentaries may provide the two characteristics of paradigms. The design profession can also adopt viewpoints which satisfy the requirements of paradigms and provide the decision environment that, in the end, may exclude realistic considerations. An example of this situation is the design approach to suspension bridges in the country over a 60 year period that lead to flexible and dynamically unstable structures. This example is explored in this note. The conclusion is that unless the second requirement is satisfied, adherence to incomplete paradigms may lead to disastrous consequences.

## History

The plot in Fig. 1 shows the ratio of L , the distance

------------------------------

*Professor of Civil Engineering, University of Washington, Seattle, WA 98195
**These requirements are parallel to the shared characteristics of paradigms of science in Kuhn's monograph (1962).

between towers, to d, the depth of the stiffening system, which is used for selected suspension bridges, compared to the date of construction. Interest has been focused on the period since the erection of the Brooklyn Bridge. Over the distance L, the bridge can be modelled as an indeterminate structure with one redundancy. The structural members are the cables and the beam of depth, d, which is considered as a measure of beam stiffness. The sag ratios are close to 1/10 for all of the bridges.

The information presented above is historical evidence from which the following trends are observable:

a) That ratios L/d of 60 to 80 were used extensively.

b) That ratios increased in the 1930's and culminated in the dimension of the first Tacoma Narrows Bridges. After 1940 the 60-80 ratios were re-established.

c) That ratios increased again with the construction of the Severn Bridge in 1966 and the use of aerodynamically designed boxed beams, as opposed to the earlier trusses and girders.

Table 1 shows selected bridges plotted in Fig. 1, where those that failed are shown with a cross and those that stiffened after 1940 by a full circle.

Table 1. Selected Suspension Bridges with Construction Dates and L/d Information

| BRIDGE | BUILT | SPAN (FT) | DEPTH (FT) | L/d |
|---|---|---|---|---|
| Brooklyn | 1882 | 1,595. | 17.0 | 94 |
| Manhattan | 1909 | 1,470. | 24.5 | 60 |
| Delaware River | 1926 | 1,750. | 27.8 | 63 |
| Ambassador | 1930 | 1,850. | 22.0 | 84 |
| George Washington | 1931 | 3,500 | 26.6 | 132 |
| Triborough | 1936 | 1,380 | 20.0 | 69 |
| Golden Gate | 1937 | 4,200 | 25.0 | 168 |
| Thousand Island | 1938 | 800 | 6.0 | 133 |
| Deer Isle | 1939 | 1,080. | 6.5 | 165 |
| Bronx-Whitestone | 1939 | 2,300 | 11.0 | 209 |
| Tacoma Narrows | 1940 | 2,800 | 8.0 | 350 |
| Delaware Memorial | 1951 | 3,460 | 20.0 | 173 |
| Mackinac Straits | 1960 | 3,800. | 38.0 | 100 |
| Forth Road Bridge,U.K. | 1963 | 3,300. | 27.5 | 120 |
| Verranzano-Narrows | 1964 | 4,260. | 28.0 | 152 |
| Severn, U.K. | 1966 | 3,240. | 10.0 | 324 |
| Humber, U.K. | 1981 | 4,626. | 14.7 | 313 |

### The Paradigm

The literature prior to the collapse of the Tacoma Narrows Bridge concentrated on theoretical treatments to predict the structural behavior of the cable-beam system. Improvements of the "deflection" and "trigometric series" methods over the "elastic" method took into account cable shape changes under live load and the consequent changes in moments and shears in the beam. The considerations were for static gravity loading and the wind loads were resisted by the bending of the deck-beam system about an axis normal to the deck and between the supports at the towers. The perceived paradigm was that:

The structural problem for the cable-beam system is to determine the most economical, functional and safe arrangement under static, gravity loads.

The conclusion was expressed in 1939 as:

". . .the most economical stiffening truss is no stiffening truss whatever; but. . .the deflections of an unstiffened truss may be too great, and that, therefore, a stiffening truss may be needed. The maximum grade change is probably of more importance than the maximum deflection;"

This outstanding paper (Hardesty 1939) was discussed by most of the leading suspension bridge designers in the country at the time and functional importance of the truss with respect to the deck geometry was re-iterated. The consequence of the paradigm was the amplified L/d ratios which are shown Fig. 1.

### The Aftermath

Fig. 1 shows the immediate return to stiff beam systems in bridges built after the November 7, 1940 failure of the Tacoma Narrows Bridge. Additionally, many of the existing bridges with L/d over 150 were stiffened so as to meet a new paradigm, even though they apparently satisfied the one previously stated. The literature after the failure abounds in papers on the dynamics of suspension bridges whereas, there was a dearth of such papers before the event. Historical studies were made and Reid's dramatic verbal record of the failure of the Brighton Pier was repeated over and over again. Even the obscure 1841 paper by John Scott Russell in the Transactions of the Royal Scottish Society of Arts in which he considered "... the general nature of vibrations that destroy suspension bridges..." was dragged out with the implication that bridge designers in the 1930's should have known better.

The profession of suspension bridge designers abandoned the previous paradigm and quickly developed novel structural arrangements which allowed even higher L/d ratios than the ill-fated Tacoma Narrows Bridge (Fig. 1). These arrangements arose from extensive wind tunnel work on Severn Bridge models which were made because of the availability of sufficient money to study the crossing but insufficient funding to build the bridge. Such research money for the consideration of various and novel alternatives is unusual; most bridges are funded, designed and constructed in the shortest period possible. Our concern here is not with response time of a profession but the apparent ease with which it can be lured by a paradigm into progressively less satisfactory structures. The difficulty was not in the understanding of the theory of Russell and others but in a constraining philosophy which was too definite to accommodate the second requirement of good paradigms stated in the Introduction, namely sufficient vagueness and ambiguity to allow the paradigms to be open-ended.

Discussion

Reasons for the implicit adoption of the suspension bridge paradigm have been suggested. An attractive viewpoint (Sibly & Walker 1977) is that wind failures abounded in the period of 1810 to 1870, but with the focus on railroad bridges the beams were made stiff to control the deck geometry. The consequence was L/d ratios of about 60 to 80. These were maintained in the post-railroad age until the glut of highway suspension bridges built after 1920. In the 1920's the "elastic" theory was abandoned and the use of "deflection" theories ensured confidence in the predictions of beam moments. When these theoretical predictions were applied to the paradigm, justification for increasing the ratios into the range which existed in the 1810-1870 period was made. An open-ended characteristic in the paradigm could have ensured consideration of dynamic effects which were being exhibited in constructed bridges. However, confidence in the calculations prevailed and the paradigm was maintained with disastrous results. It is possible to see a similar situation today where impressive analysis is the dominant evidence in describing the behavior of designed structures. If the environmental circumstances have short return periods then the design profession can adjust the theory in the light of observations (as could have been the case in the late 1930's) and review the paradigm. The difficulty must arise in the prediction of behavior in the light of circumstances with long return periods. It is possible that confidence in calculations may lead to catastrophic conclusions. One hope is that sufficient laboratory and full scale testing will accompany the calculations to justify the paradigm. Certainly the increase in the L/d ratios after 1966 have been justified by a mixture of theory and wind tunnel tests.

Conclusion

Errors by a segment of a profession can be made. The

consequences are far more widespread than an error made by an individual. The modus operandi can be expressed as paradigms which are adhered to by the profession and vague enough that validity is maintained. It is the last point of vagueness, with the associated open-ended character in the paradigm, that prevents it from becoming a modus vivendi.

Acknowledgement

This work was supported by the National Science Foundation under Grand No. ECE-8518155.

References

Kuhn, T. S., The Structure of Scientific Revolutions, University of Chicago Press, 1962.

Hardesty, S. and H. E. Wessman, "Preliminary Design of Suspension Bridges," Transactions, American Society of Civil Engineers, Paper No. 2029, Vol. 104, pp. 579-608, Discussion pp. 609-634, 1939.

Sibly, P. S. and A. G. Walker, "Structural Accidents and Their Causes," Proceedings, Institution of Civil Engineers, Part 1, Vol. 62, pp. 191-208, 1977.

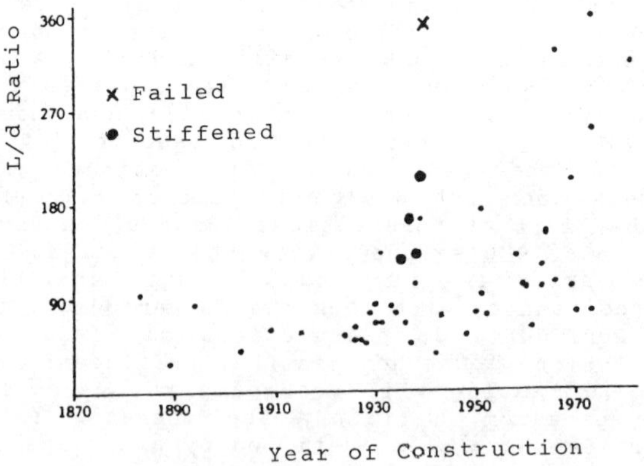


Fig. 1. - Plot of L/d Ratio Against Construction Date

# WHERE SHOULD WE LOOK FOR ERROR CONTROL

Owen G. Ingles*

Abstract

This paper considers briefly the various means by which error control is currently accomplished. Many of these methods are indicated to have shortcomings which probably impede any further useful refinement of them. It is stressed that what is needed are not merely measures against human error, but economical ones. Better communications, not only within the engineering profession but also with the community at large are considered essential to improved error control; whilst this is tackled, the engineer should stress the immediate usefulness of monitoring systems. It is argued that "human error" has been redefined legally to the engineer's disadvantage, and this situation also needs urgent attention.

## Where Should We Look for Error Control

There is now abundant evidence that man's mechanical, computational, and measurement skills (inter alia) have far outstripped his proneness to error. The fallibilities of the "human machine" are controlling its performance. Our future advancement depends on understanding and overcoming (or at least circumventing) human error.

To cite but a few examples: Matousek & Schneider (1976) found upwards of 85% of building failures to be sourced in human error: Melchers & Harrington (1982) found an error rate of 1% for simple design tasks: Ingles (1984) showed failure rates in Australian dams to be 0.2% per structure per annum - subsequently upgraded to 0.3% - an incidence surely not envisaged by their designers! Theoretical expectations are clearly very different from practical reality.

The legal view of human error incidence is similar to the engineer's (above). A short analysis of the findings of Royal Commissions, Select Committees, and Boards of Inquiry into Road, Rail and Mining disasters in Australia since 1858 shows that 53% were directly attributed to human error (some described as "gross negligence") and a further 20% to possible human error. Less than 7% were attributed to chance (i.e. random events, "Acts of God", etc.).

---

*Risk Management Consultants, Swan Point 7251, Tasmania, Australia

It should not be unexpected that "failure" (however it is defined) has a high, often dominant, component of human error because, in any chain of antecedent events, the individual links are either human or traceable to human intervention at some earlier time. In a strict sense, all except the unforeseeable is human error. What might be termed as "Acts of God" have almost disappeared beneath the forward thrust of technology (volcanic eruption, meteor impact, and a few similar events are all that remain).

Random change of floods, earthquakes and other natural phenomena are now made foreseeable by means of "return periods", statistical or probabilistic, placing a human error element on their damaging occurrence.

It hardly needs to be stressed in the United States how legal judgements, in the wake of failures, attribute personal negligence as the major cause. This tendency has also risen sharply in other countries, such as Australia.

Because virtually all performance not in accordance with expectation (one quantifiable definition of failure) is now attributed to human error, small improvements will not be sufficient. We need control methods which will reduce error not by one, but by two - or more - orders of magnitude. Maybe this will involve simultaneous use of more than one method. It will be the object of this paper to discuss some of the possibilities.

Consider first the error-proneness of the human "machine". It is far too high by comparison with purely mechanical system components. As with other machines, the error rate can be reduced by proper maintenance and operation. This is the reasonably well researched field of ergonomics. We know, for example, that error is reduced by optimum room temperature, rate of work, noise type and level, etc. (cf. Ingles 1979) - but we already operate reasonably efficiently with respect to working conditions, so that gains in this area would be small, except insofar as we could eliminate distractions. (These don't necessarily come in skirts; there are many serious psychological ones such as worry, fear, etc. with far greater importance). The difficulties in reducing these kinds of human machine errors are immense because every "machine" is unique, so that uniform prescriptions will have only limited validity. There is room for more research, but the probable outcome seems incommensurate with the effort required. In short, we cannot yet create Superman: and would do better to restructure our systems to be Foolproof.

Various methods have been proposed or tried for overcoming these limitations of the human machine. An obvious one would be to develop selection processes whereby, only those persons found to be or trained to be least prone to error were made responsible for important works. In effect, we adopt this method when we insist on "professional qualifications" for our engineers. Our concern today about the need to reduce human error is clear evidence that this method is inadequate (insofar as it is presently practised). It is my belief that engineering education can be substantially upgraded.

Another popular method in use is checking, based on the assumption that different persons have different error patterns, so that superposition will largely eliminate errors. The research of Harris & Chaney (1969) suggests strongly that this is illusory, even as much as 10% of critical defects remain undetected after 6 independent inspections. It is a worthwhile method, but not sufficient as it is presently practised.

Safety Factors are the time-honored method used by engineers to cover their fallibility and/or inadequate knowledge. As Lind (1979) suggests, present safety factors are probably sufficient to cope with one error in the system, but not two or more. Nowadays we have good probabilistic analyses which relate the risk of failure to the safety factor(s) adopted, (e.g. Lumb's (1974) treatment of slope stability), so that a safety factor can be chosen if we have agreed on a desirable level of risk. Acceptable risk has been extensively studied in recent years, as exemplified in the 1981 book of that name by Fischhoff and coauthors and in the civil engineering field by Ingles' (1982,1983,1985) studies, but very substantial problems remain in the interpretation of acceptability. For one thing, analysis has hitherto been almost wholly based in fatality statistics, because the quantification of suffering resulting from injury to persons as opposed to things is too intractable. Cost-benefit studies can be used to relate the safety factors with their associated risks, so that an optimal cost benefit is achieved which is critically reliant on the assignment of a "value of life". The latter concept can provoke such strong moral reaction that it is specifically excluded from some safety treatments such as the Swiss SIA 260 for structures (1980) which reads:

Personenschadenrisiker ..... werden im Rahmen der Anwendung des Normenwerks des SIA grundsatzlich nicht Geldeirheiten gegenuber- gestellt und sind damit im Einzelfall okonomischen Ueberlegungen entzogen. (Das heisst, eine Umrechnung von Merschenleber in Schweizerfranken ist im Normenwerk des STA nicht zugelasser).

As Ingles (1983) has argued, we are already working with safety factors which produce results very close to community tolerance (acceptability) levels. Probably this has been forced by economic pressures. It is easy to increase the safety, but not at a low price. One matter which must be urgently addressed is, how much is the community prepared to pay for increased safety measures? This is the ultimate determinant of what measures we can adopt. We must be vitally concerned with what are the most economical measures against human error. So it may well be that safety factors are not a good means for error reduction.

Redundancy has often been applied, especially in structures, to overcome the dangers of error. It can be effective, but expensive. There may well be merit in extending studies on the reliability of system components.

Codes of practice have been successful in reducing human error, as seen in the results in Ingles & Nawar's (1984) study of domestic housing failures, for example. Codes of practice minimize human error by expressing the summative wisdom of a large body of responsible engineers. Their cost is usually expressed in overdesign and the less tangible losses of innovation.

Many other methods, either proposed or in present use would appear to have no significant advantage over those described above; with one possible exception, motivation. Is it possible to stimulate the human machine to better performance? Throughout historical time, penalties and reward have been used for this end. When the Persians sought to cross the Hellespont into Greece, the bridge engineer who failed was promptly executed whereas the engineer who succeeded was rewarded with vast wealth. The rate of bridge failures at that time is not recorded, but probably remained unchanged. Engineers today do not see either penalty or reward as important to the error rate (cf. Ingles & Nawar, 1984), though this perception is clearly changing under the pressure of adverse legal judgments in professional liability suits. So, it may be appropriate to reexamine the motivation path to error reduction; though this work may need to be done by psychologists rather than by engineers.

Ingles' survey of Australian engineer's attitudes showed that the greatest reliance was placed on clear instructions and operational simplicity if error was to be minimized. This is only another way of saying "do not exceed the complexity tolerance of the human machine", a point clearly made by Harris & Chaney in 1969. Systems complexity is increasingly difficult to avoid, so we must fall back on improved communication. And that encompasses education too.

Finally, remembering that cost will exclude many methods for error reduction, perhaps a better approach would be to continue as at present (nil cost) but devise and insist upon monitoring systems for all engineering works during their service life. The objective would then be to establish early warning systems of negligible cost compared to the structures themselves (it is presumed that something could be done to save the structure in the time allowed by the warning). Again, reliability theory is essential to the successful operation of such monitors.

So, where do we stand? It seems unlikely that we can achieve large reductions of the error rate except at large cost or after further extended research. For immediate purposes, it would seem better to acknowledge that risk reduction will not be achieved by objective methods because it is value laden, and therefore the most pressing need is to reach agreement about certain crucial concepts and their definitions. Quoting Fischhoff et alia: "Beliefs about the facts of the matter shape our values; those values in turn shape the facts we search for and how we interpret what we find." We need measures of community beliefs and values, measures which are sensitive to their dynamic changes, agreement on how they should be used (the mean, the median, some percentile, etc.).

One example is the concept of error criticality. Some errors are worse in their effect than others. If we could concentrate on removing those errors which lead to total system loss, and pay less attention to minor errors which might be easily corrected as and when they appeared, then our process of error reduction might become far more cost efficient. There is, as yet, no agreement on what constitutes a critical risk. (Pickrel & McDonald in 1964 defined "catastrophic" errors as those which produce severe degradation - loss of system, or death, or multiple injuries). Critical risk probably depends on the number of people at risk (cf. my own finding (1983) of a strong logarithmic relation between the number at risk and the acceptability of the risk). Since type and magnitude of risk are infinitely variable, we almost certainly need a number of defined points - in the same way that geomechanical engineers have defined various particle size points as boundaries between clay, silt, sand and gravel. The goal is to communicate clearly to others, the level at which we have analyzed each particular problem. There are other value judgements involved, too. Errors will be "de-rated" if their effects are accepted voluntarily, which can be achieved by superimposing appropriate rewards. Motivation, referred to earlier, is essentially a value judgement.

Other values requiring urgent clarity by definition are "failure" and "quality control", inter alia. It was notable and regrettable that IABSE in 1983 was unable to reach any agreed position on the latter concept. Surely an attempt should be made by an International Committee to publish provisional definitions so that debate could begin and mutual understanding enhanced.

Some aspects of the criticality problem can be circumvented by sensitivity analysis of the kind recommended by Nowak (1983). Sensitivity analysis enables us to identify quite objectively those aspects of the total system most likely to lead to system degradation, and hence, should receive priority attention. But it says nothing about the effects of that degradation, so is incomplete per se.

As Australian surveys show (Ingles 1983a & b), engineers place great weight on experience for error reduction. It might therefore be argued that engineering education requires more emphasis on practice, even at the expense of theory. As remarked before, the actual failure rates seem to exceed by a considerable margin what we expect from our theory-based designs. One significant omission in engineering education are courses in the analysis of failure. The old saying that "we learn by our mistakes" surely applies. One reason for this omission is, of course, that the literature on failures is sparse. We have pre-ferred to write about our successes. Students today have an unwarranted confidence in the design methods they were taught, and must be objectively disabused of this. One important field for immediate research and teaching, if human error is to be reduced, must be (as implied at the outset of this lecture) a detailed study of the findings of Commissions of Inquiry into engineering "failures", "disasters", "catastrophes", etc. Such a study must also consider the arrow of time. This approach seems a necessary prerequisite to the quantitative formulation of error reduction methods.

What can be done pending the results of such research? We must continue to insure adequately; remembering that insurance will not be purchaseable unless we can demonstrate that we have proper loss-prevention methods in place, including for instance checklists for staff, clear delineations of authority, and plans which minimize the risk exposure (e.g. don't plan to build bridge foundations during the peak flood season). Insurance is a first class tool for risk management, not because it can reduce human error (some might argue that it does the reverse), but because it can be an accurate measure of the error control which we have achieved. Success in error control should be directly reflected in premiums paid for insurance in the same manner as accident-proneness increases our auto insurance.

## Conclusions

Human error is always controlled by community tolerance levels, both historically and today. Being thus value-laden, there are two prerequisites for its controlled reduction; the formulation of agreed value definitions, and their communication to the community at large. The community must be involved in engineering decisions of this kind, and we must be extremely cautious about proposing any numerical methods whatsoever until that has been done. Consider the situation where, if structure A failed 40 people would die, and if structure B failed there would be a one-third probability of no deaths and two-thirds probability of 60 deaths. Which is the preferable structure? Whilst such definitions are hammered out, having particular regard to the equitable disposal of finite community financial resources (the weighing of cost and benefits seems unavoidable), education should be focussed on the analysis of failure so as to sharpen awareness of the human error problem and to uncover better means for combating it.

In the meantime, we have no choice but to continue with existing measures such as safety factors, checking, selection, codes, monitors, penalties and rewards, etc. even though recognizing their insufficiencies in scope, cost or effectiveness. Most of them can be be marginally improved thanks to present mathematical advances in probability theory and reliability theory. (Fuzzy sets are also perhaps of value insofar as discrimination errors are important). Some key diagrams from earlier sources are appended for illustration of the arguments presented here (Figures 1-5); but chapters 9 and 10 from Fischhoff et al. (loc.cit.) must be recommended reading for the topic. A particularly detailed treatment of human error in engineering is also given in Ingles (1985), to which reference should be made for its historical and societal aspects.

References

Fischhoff,B., Lichtenstein, S., Solvic, P., Derby, S.L., & Keeney, R.L. (1981), "Acceptable Risk", Cambridge University Press.

Harris, D.H. & Chaney, F.B. (1969), "Human Factors in Quality Assurance", Wiley.

IABSE (1983), "Workshop Rigi 1983".

Ingles, O.G. (1979), "Human Factors and Error in Civil Engineering", Proc. 3rd ICASP, Sydney, III, 402-417.

Ingles, O.G. (1982), "The Perception of Risk by Australian Civil Engineers, and its Implications in Design", Proc. Symp. Risk Assessment in Geomechanics, Inst. Engrs. Aust. (Old. Branch).

Ingles, O.G. (1982), "Measurements of Risk and Rationality in Civil Engineering", Proc. 4th ICASP, Florence, I. 357-373.

Ingles, O.G. & Nawar, G. (1983), "Evaluation of Engineering Practice in Australia", Proc. IABSE Workshop Rigi 1983, 111-116.

Ingles, O.G. (1984), "A short study of dam failures in Australia, 1857-1983" Civil Engg. Systems, 1(4), 190-194.

Ingles, O.G., (1985), "Human Error, and Its Role in the Philosophy of Engineering", Doctoral Thesis, University of N.S.W., Australia.

Lind, N.C. (1979), "Optimization, Cost-Benefit Analysis, Specifications", Proc. 3rd ICASP, Sydney, III, 373-384.

Lumb, P. (1968), "Statistical Aspects of Field Tests", Proc. 4th Conf. A.R.R.B., Melbourne, 4(2), 1761-1771.

Lumb, P. (1974), "Application of Statistics in Soil Mechanics", in "Soil Mechanics, New Horizons" (Ed. Lee), Butterworths, pp. 44-111.

Matousek, M. & Schneider, J. (1976), "Utersuchungen zur Struktur des Sicherheitsproblems bei Bauwerken", Bericht Nr. 59, ETH, Zuerich.

Melchers, R.R. & Harrington, M.V. (1982),"Human Error in Simple Design Tasks", Monash Univ., Civil Engg. Rept. No. 3.

Nowak, A.S. (1983), "Strategies for Control of Human Errors", Proc. IABSE Workshop Rigi 1983, 169-172.

Pickrel, E.W. & Mc Donald, T.A. (1964), "Quantification of Human Performance in Large Complex Systems", Human Factors, 6(6), 647-663.

SIA (1980), "Weisung fur die Koordination des Normenwerks des SIA in Hinblick auf Sicherheit und Gebrauchsfahigkeit von Tragwerken", 4 Fassung SIA 260.

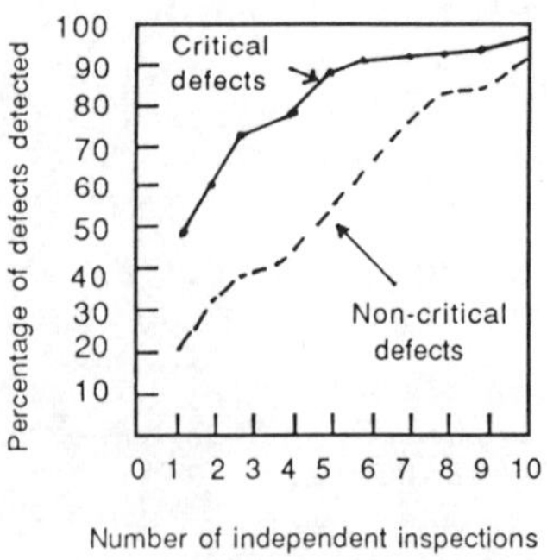


Fig. 1 Error Reduction by Independent Inspections (after Harris & Chaney 1969)

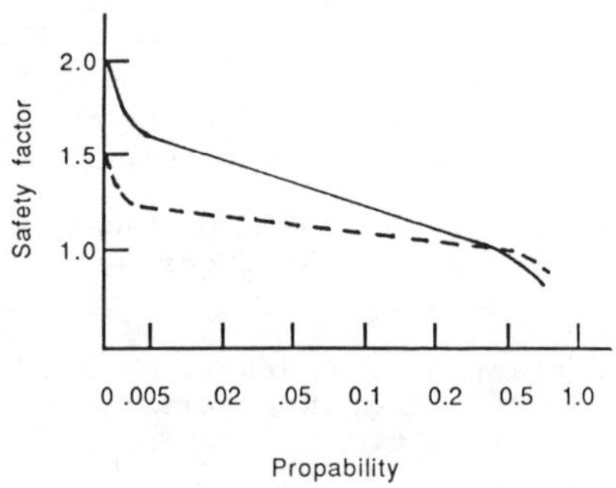


Fig. 2 Probability bounds for slope stability (Lumb 1974)

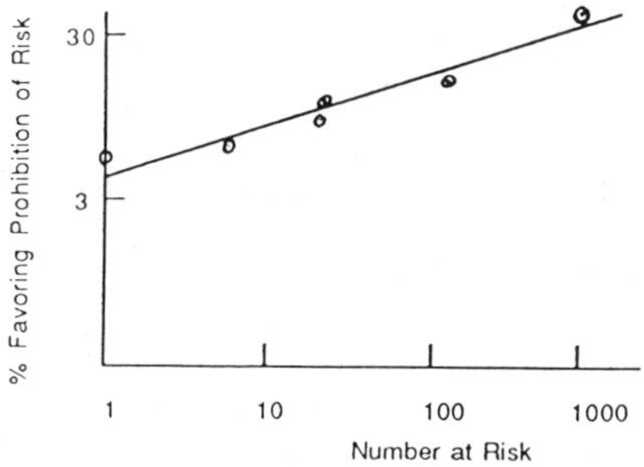


Fig. 3 Aversion to Disaster (a polled acceptability of risk, fatality based) (after Ingles 1983).

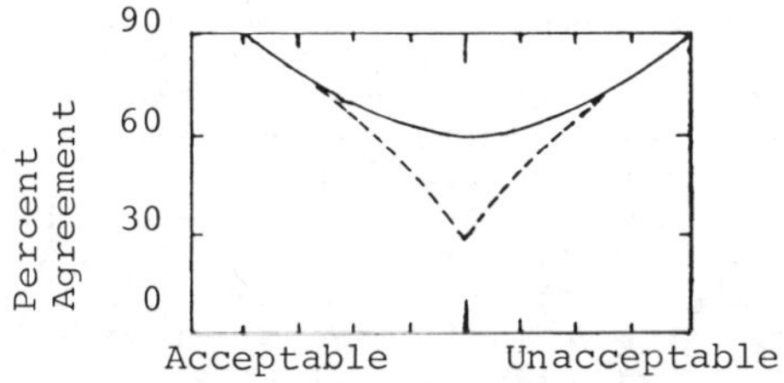


Fig. 4 Effect of Boundaries on Error Rate (the problem of definition and recognition of boundaries) (after Harris and Chaney 1969).

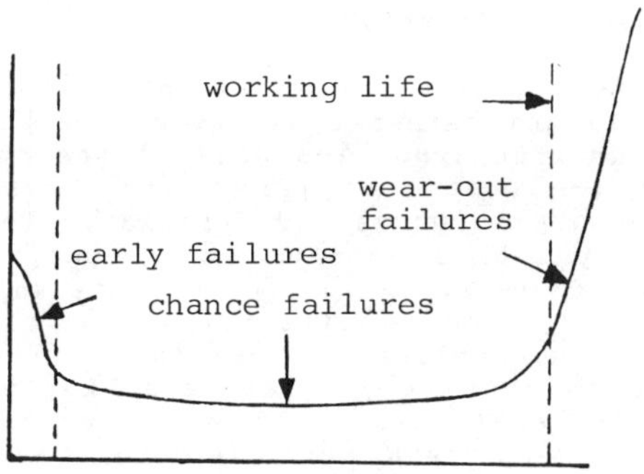


Fig. 5 Failure Rate Diagram (the occurrence of failures, showing the most critical periods for attention) (after Lumb 1968).

# H.E. IN THE DEVELOPMENT OF STRUCTURAL SPECIFICATIONS

T. V. Galambos*, M. ASCE

## Abstract

Errors in the use and in the development of structural design specifications are discussed and existing measures to eliminate them are enumerated.

## Introduction

Structural Specifications are codified rules for design criteria, which must be adhered to in the design and construction of buildings, bridges, industrial plants, etc. so that the final structure is both "safe" and "econcmical". The emphasis in such codes is on safety, but, as a practical measure, specifications which do not let the designer create also a serviceable and economic structure, are useless. Structural specifications are developed and maintained by a variety of interlocking committees, and so there is always a possibility of human error. Such errors can be quite far-reaching in their consequences since the specifications are used by many engineers in a particular industry, say the steel construction industry, and thus a whole class of structures can be affected.

Human error can originate in a) the improper use of a specification and b) in the development process. The following comments will address these two issues.

## Improper Use of a Specification

Structural specifications are compact documents setting forth a linguistic and mathematical terms the minimum standards for structural design. The body of the specification is written, for the most part, in a very terse language, using extreme economy in words and formulas. The main text is supplemented by tables, charts and graphs, and further explanatory material is given in a Commentary which attempts to explain the background and the derivation of the rules, and which lists the pertinent research references. These explanations are at best very brief, and they can be properly understood by people with previous experience and education in the subject matter of structural analysis and design. The detailed exposition of structural specifications

---

*Professor of Structural Engineering, University of Minnesota, Minneapolis, MN 55455

is provided by textbooks used in courses of structural design, and it is implicitly assumed that code use is practiced only by people having previous educations in structural analysis and design.

A very large share of human errors result from the design application of structural specifications by persons having only an incomplete or vague understanding of their meaning. Even though many examples of malfunctions of structures could be cited, this will not be done because the point is obvious. Control, in the form of the review of designs by professional registered engineers, should be insisted on.

Practicing experienced professional engineers may have a wide spectrum of understanding about what structural specifications are supposed to represent. These views vary, on the one hand, from considering the specifications to be "minimum standards" to be used as "guide-lines" in design together with "judgement", "experience" and structural analysis extending beyond the explicit rules, to, on the other hand, an interpretation which thinks of them as inflexible legal standards which must be obey ed at all times. Either of these views can result in human error, and in cases of ambiguity mature designers will draw on outside expertise which is generally always available from the people responsible for maintaining the particular specification. However, errors resulting from the misunderstanding of the rules, or from using them beyond their intended context, abound in practice. They can be avoided by continued learning from seminars and short courses on the subject matter of the code.

## Development of Structural Specifications

The United States is unique among the world's industrialized nations in that its structural design codes are developed as voluntary consensus documents prepared by the individuals and groups affected by its use, not by governmental or semi-governmental agencies. As a result, there is not one structural design code, but a profusion of specifications serving different and often overlapping constituencies. To illustrate this point only a few examples will suffice. Steel building structures fabricated from hot-rolled steel plates and shapes are designed according to the specification of the American Institute of Steel Construction (AISC), while structures made from cold-formed steel are governed by a specification of the American Iron and Steel Institute (AISI). Bridges are under the codes of the American Association of State Highway and Transportation Officials (AASHTO) if they are highway bridges, and the American Railroad Engineering Association (AREA) if they are railway bridges. Both of these bridge specifications cover steel, concrete, wood, masonry and aluminum bridges, while concrete buildings are designed by the American Concrete

Institute (ACI) Code.

These structural design codes overlap with each other to some extent (e.g. frame stability criteria in the ACI and the AISC Specifications) while otherwise identical situations are handled entirely differently in different specifications (e.g., shear panel design by buckling theory in the AREA Specification and by tension field theory in AASHTO and AISC Specifications). Furthermore, these design codes make reference to material specifications (American Society for Testing Materials), welding codes (American Welding Society), bolt codes (Research Council on Riveted and Bolted Structural Joints) and load codes (American National Standards Institute), to mention only some instances of interdependence.

Each of these codes, standards and specifications are developed, expanded, improved and up-dated by individual advisory groups consisting of voluntary unpaid experts representing users, producers, designers and researchers. The final responsibility for approval rests then with a particular association, such as AISC, AISI, ACI, AASHTO, AREA, etc. where another set of actors takes the work of the experts from the advisory groups and integrates it with the specific aims of the Association. In the final outcome the codes are then adopted as written, or modified for local use, by regional Building Codes where they eventually take the form of a legally binding document.

There are thus virtually hundreds of individuals and scores of organizations which interact to produce the dozen or so structural specifications which must be considered by structural designers in their daily task. On the surface of it this situation appears to invite human error on a grand scale. Attempts at standardization of nomenclature, of code rationalization, of integration between all codes, and of unification in national specification system have so far proven to be unsuccessful, and will remain so in the future. Surprisingly, chaotic as it may seem, the system works quite well.

Following are some important human error scenarios which are based on actual occurrences which were corrected during the development of the AISC's Load and Resistance Factor Design (LRFD): (This new specification is due to be issued in late 1986. Its research and development spanned the period 1969-1986, and it involved the collaboration of some one-hundred individuals, including an advisory group of nearly 40 experts.)

1) Important omissions of key provisions, especially
   a) at interfaces between various parts of the specification,
   b) during revision cycles.

   These errors are best combatted by using decision tables

and flow charts.

2) Ambiguous wording of rules.
3) Changes in the definition of symbols between different parts.
4) Incompletely tested concepts for provisions which never appeared in the parent AISC Allowable Stress Design (ASD) specification.
5) Incorrect translation of rules from the ASD to the LRFD specification.
6) Changes in related specifications (e.g. the bolt specifications) are misunderstood or incompletely applied.
7) Communications failures between some of the involved actors.
8) Lack of agreement on technical issues between members of the various committees (e.g., inability to define a basis for a column formula).

Summary

It is impossible to create a set of error-free specifications which are, in addition, also incapable of being misunderstood or misused. However, it is possible to avoid most errors by involving only conscientious and competent professionals in both the development and the use of structural design codes. The great utility and success of such codes in use today attests to the fact that the building professions have learned to cope with the aspect of the human error problem.

## CHECKING TECHNIQUES

Franz Knoll*

Abstract

The paper deals with the principles of existing checking techniques. Since no scientifically consistent method exists to determine effective strategies, common-sense must step in and does so, in the professional practice, permitting some seemingly trivial but nevertheless pertinent conclusions. Some guidance can be derived for the planning and operation of checking activities.

### Introduction

In the absence of consistent scientific knowledge about gross errors, their consequences and their elimination, the practice is presently - as always in situations of insufficient knowledge - filling in with intuitively guided activities directed towards the discovery and correction of errors.

While we have generally come to recognize the fact that usually when things are going wrong, there has been some human involvement generating havoc in the building process which is otherwise assumed to be smooth, orderly and predictable, opinions about the type, amplitude and timing of remedial action are still widely divergent, ranging from formalized procedures involving great amounts of man-hours and paperwork, to targeted intervention of highly qualified individuals, with any description possible between the two extremes. All these opinions have also been applied in various circumstances, with results which are in general being found adequate but without any objective way to judge whether the method applied is the method of choice in reality and not just in the minds of those who adopted it. Criteria for this choice can be found deriving from economics:

What is the cost of the error-hunting?

to legal aspects:

---

*Nicolet, Chartrand, Knoll Ltd., Montreal, Quebec H3B 1N7 Canada.

What am I going to tell the judge when something goes wrong?

or the expedience of the day:

Whom could I send today to look at this reinforcing steel?, etc.

What is rarely considered for the choice of method, is the real effectiveness of the procedure to be adopted. This is clearly because so very little is known about this particular property. Beyond a rather fuzzy consensus of a number of experienced people, regarding some purely qualitative relationships (see e.g. Report on Rigi Conference, Schneider 1983) everyone is left to his own wits when it comes to make sure about the quality, i.e. the "rightness" of what is being produced, through eliminating the "wrongness" whenever it, unpredictably, may turn up.

On one hand therefore, the general record of engineers has been rather satisfactory on the whole, the sum-total of "wrongness" in the building trade being what society wants - or accepts. On the other hand, the loss through the lack of quality has been and still is, an important factor in the economy, amounting to something around 5% of the total expenditure. Also, society does take rather dramatic exception to individual mishaps and punishes individuals found to have been wanting, or merely involved, rather severely.

This apparent discrepancy or rather "double-standard" can only be explained by looking at the situation of the responsible individual: He will know perfectly well that if something turns sour under his hands, all the general acceptance of society regarding the rate of mishaps, will not help him personally. It has been stated by many people in various circumstances that the prospect of discomfort implied by the consequences of an error, is still providing the main - albeit a negative - incentive to the individual involved in the building trade, for watching his every step and decision. Seemingly then, this incentive represents the way society - consciously or not - tunes conditions to its collective liking: Not by creating rulebooks or safety factors but by threatening the human individual. The modern insurance mechanism was thought to fulfill a mitigating function in this situation, by softening at least the financial blow an individual - engineer, architect, contractor, superintendent - would suffer, by spreading the consequences of liability onto everybody. However, insurance in construction industry does not appear to be in a good way presently and it may well have to change soon into something less than the general protective umbrella we liked to be used to.

What is the answer of the individual to this dilemma?

The Checking of One's Own Work

Only fools can believe they do not make mistakes. Everyone else must reckon with the situation where, through lack of attention, overwork, too much or not enough pressure, etc, etc, some of his actions/decisions are taking a wrong turn. Some evidence was found that in the first round of design work the rate of error is terrifyingly high, i.e. several orders of magnitude greater than at the end of the game, which can be reasonably chosen at the beginning, or even at the end of the usage of a structure, Melchers et al (1984, 1985).

What happened in between, we all know, was the elimination of most of the errors through correction: Wrong numbers were recognized as wrong and replaced by right ones, bad decisions were found out and rectified, inadequate designs were improved. Obviously, this process of elimination of errors must be quite effective in reality and it is thanks to it that a generally satisfactory equilibrium is maintained between the current rate of "wrongness" and the additional expenditure which would be required to reduce "wrong" in order to augment "right".

The individual having produced a decision, i.e. an element of information, concerning a structure to be, can do a number of different things subsequently:

- He can believe that his decision was right. The consequences of this is trivially that he leaves it at that, proceeding to other activities.

- He can suspect that there might be something less than right to his decisions. Consequently he will now have to find ways to, either assure himself that, after all he had been right, or, in case he can identify something being wrong, correct the decision accordingly. In both cases he will have to proceed to the activity of checking, i.e. verification.

- In the simplest case, he will do the checking himself, e.g. going once more over the calculations of the day before.

- For various reasons he may decide to delegate the checking to someone else: a subordinate, a peer, an institution, or another party in the construction process. Each possibility has, of course, its own special implications which in many ways will influence the choice. Peer review will likely result in a more or less friendly discussion, with a compromise over some disputed aspects, reached in the common interest of both individuals. The design review by institutionalized bodies (bureaux de controle, Prufingenieur, etc) usually results in conservatism, the particular personal concerns of two or more parties having to be satisfied. The findings of errors by other parties in the building process, on the other hand, will likely result in adversary situations, with that other party attempting to exploit the discovery for its own ends, etc.

However, none of these side-effects has anything to do with the real question, and presently no knowledge exists beyond the limited personal experience of seasoned individuals, as to the effectiveness of various methods in hunting down gross errors.

For the discussion of checking techniques a variation of the filter model may serve as an illustration:

The input consists of a liquid with a variety of ingredients in suspension, solution, or of solid form. The target is a liquid with no solids of a size greater than what is covered by reasonable safety factors. The model also includes vessels the faces of which are not all or not always transparent, and people handling (and observing) the liquid, i.e. casting it from one vessel into another, passing it through filters of a variety of characteristics, some of which are open to choice. Let the structural engineer be generally in charge of the filtering process. However, his time and the remuneration of the personel he may employ, are limited. Also the people working on it are not all motivated purely towards the target but have their own exploits to follow, as well as axes to grind. To further adapt (complicate) the model, let us assume that the processing includes addition of further ingredients, some of which will react with those already present in the liquid, e.g. in solution, and produce new solids. The liquid itself is also more or less opaque, the degree of opaqueness representing for instance the degree of complexity of the design.

The engineer as the chief tactician, is now in charge of managing the filtration process, and at the same time, optimizing it. He will have to consider the condition of the filters, and the condition of the liquid when it is passed through each particular filter, and all together his efforts have to stay within the total means allotted to the purpose. Some of the filters will be applied anyway, e.g. the contractor will scrutinize the drawings for mistakes, if it is only to find ways to claim extra cost. Other filters come out of the engineers' arsenal and he may apply them when and where he decides, within some limitations.

Most filters are directional, i.e. they will only catch certain classes of errors: A construction supervisor with a long experience but with limited education will see anything unusual and - hopefully - report it in time. He will however, not be able to be critical in the face of a new concept. A draftsman checking shop drawings will notice discrepancies with the engineers' drawing but he will not catch a detail which does not work. Similarly, limitations for any find of filtering activity can be derived from the limitations of the people executing it, and if it's only a limitation in time when, say, the chief engineer himself goes out to look after his work or that of his subordinates.

The manager of the checking procedure must be entirely conscious of the type of errors which was likely to have been caught by each and every check, and inversely, of the kinds of errors possibly still at large. Also the filtering may in each case be less than perfect, some of the liquid spilling over unnoticed, or the filter itself having holes - people not paying enough attention to what they are doing.

What To Check? When to Check? How to Check? Who to Check?

These four questions define the problem of the managing engineer. They have in common that there is a trivial and obvious answer which comes to mind easily but which turns out to be either impossible or impractical to follow in the practice.

What to Check?

Of course: Everything or least everything that matters needs to be checked. Every decision or element of generated information must be followed and verified at every station where it is converted into a different form, context, or linked up with other elements.

This has been tried with great expenditure of effort, money and manpower, e.g. in connection with nuclear power stations, with limited success as we know today, but with the side-effect that nuclear power stations have been nearly priced out of the market.

It is of course easy and presently also popular to criticize the "nukes". However, a very important and valuable piece of information was generated in that context, namely that the mere quantity of checking effort is not directly related to its aim, the elimination of errors. If a certain level is exceeded further effort will no longer improve the situation but will generate only paperwork and make-believe. The reasons for this are common sense and can easily be understood from the situation of the checking agent whose abilities of detection degenerate after a certain amount of monotonous work.

Therefore: The optimum does not, after all, lie with the maximum effort possible but with a judicious middle of the road, and of course, other parameters need to be considered.

A promising approach appears to be in all cases to gear the checking effort to the importance of the subject, e.g. the consequence of something going wrong. However, many present-day practices are being exercised which totally ignore this exceedingly obvious and plausible notion.

For example in the case of most concrete structures, the compressive strength of the concrete is scarcely related to the performance of the structure, with the exception of lightly reinforced compression elements. This would mean that quality assurance and checking ought to be directed elsewhere - where it matters, for instance onto the correct placing and fixing of the reinforcing steel, the attachment details, or onto the parameters related to shrinkage - probably the one single property of concrete which causes the most damage. In some cases the obsession with the compressive strength can aggravate the situation rather than improve it - e.g. through the use of excessive amounts of cement just because one is nervous about the cylinder tests. This will raise the strength unnecessarily but also increase the heat of hydration and shrinkage.

In the case of the concrete structure, the optimization problem for the direction of the checking effort could be solved using a priority list, about like this:

| Subject | A<br>Likelyhood of serious error | B<br>Likely damage | C<br>Priority = compound of A & B |
|---|---|---|---|
| Misplaced reinforcement | medium to high | structural failure = very serious | high |
| Inadequate curing of concrete in the structure | high | excessive cracks, deflections, deterioration = costly | high |
| Misalignment of forms | high | not very serious can be absorbed easily in most cases | medium |
| Air content of concrete in cold climates | high | will cause frost damage = very serious for concrete exposed to exterior environment | high |

continued

| | A | B | C |
|---|---|---|---|
| Subject | Likelyhood of serious error | Likely damage | Priority = compound of A & B |
| Design of reinforcing | medium | may cause structural failure = very serious | high |
| Water access to the completed structure | high | will cause deterioration, often difficult to detect = serious | high |
| Concrete compressive strength for bending elements = beams, slabs, walls | low | low | very low |
| Concrete compressive strength for columns | low | important for highly stressed columns with low reinforcing and few stirrups | selectively high |
| Foundation design | low | important, equal settlement can cause costly damage = serious | medium |
| Design decisions made in a hurry | high | possibly serious | high |
| etc. | etc. | etc. | etc. |

Of course the priority list is open-ended and will vary with circumstances and also with the subjective perception of the individual: He will quite naturally assign higher priority to subjects which have caused him trouble in the past. However, great precision is not of essence and therefore, a little subjective bias is quite acceptable.

## When to Check

The trivial answer to this is that the check, in order to be final and all inclusive, must be performed at the end, i.e. at the final conversion of the information, e.g. on the finished structure in the form of a load test.

Quite frequently this was in effect done at substantial cost and commotion. Load testing a structure usually implies that confidence in its performance and safety is insufficient because of whatever circumstances during design and construction. In most cases it has the status of a last resort, while in other cases it may be quite impossible to perform: How would one load test a building for settlements, or a suspension bridge for wind loads?

Once more therefore, the trivial and obvious answer is not the real one, which must be sought within the limitations applying to the practical case as it were.

Two basic considerations can be stated which, together, define the conditions of the optimization problem the engineer is facing when he determines the timing of the check(s):

1. The later in the process the check is applied, the fewer "fish" can be expected to escape the net.

2. The later in the process the more costly it is likely to be to apply the filter - i.e. to correct errors.

Some corollary conditions may arise from particular circumstances such as the discrete impossibility to verify reinforcing steel after the concrete has been placed, or the finality of a differential settlement and its consequences, after it has been allowed to occur.

In practice however, a good timing of each checking activity can in most cases be selected easily or even trivially through common sense, e.g. as follows, again using the routine case of a reinforced concrete structure:

| Subject | Good or Optimal timing |
|---|---|
| Framing concept | Before detail design is begun on a large scale. However key details should have been established. |
| Details of reinforcing | First check following the production of construction drawings.<br><br>Second check when steel is in place.<br><br>Very often mistakes become apparent only at this stage, especially in complex geometri- |

| | |
|---|---|
| | cal situations. |
| Detail check of calculations | No ideal time exists for this, because:<br>- the half life of design assumptions is short;<br>- modifications keep being introduced until construction time<br>- the check does not catch mistakes of information transfer: calculations to reinforcing plans, bar lists, delivery and placing. |
| Concrete strength | Before stripping of forms for suspended elements. This is when damage is initiated.<br><br>As early as possible for compression elements, since later it will be very difficult or impossible to correct any deficiencies.<br><br>In both cases the 28-day cylinders are obviously quite useless, and other methods should be adopted. |
| Foundation conditions | Before the framing concept is established, in many cases even before the building site is selected. |
| Quality of construction work: formwork, curing, placing | Early in the construction period, to be repeated and intensified if a slackening is observed. Most projects experience a learning period for everybody at the early stages, until the process runs smoothly. Later, the setting in of routine may again cause lack of attention and errors. |
| Design strength of members | At the stage of construction drawings. In this way, most modifications and changes of circumstances will be considered. Also the best chance for misunderstandings and gross mistakes to be caught and corrected, is offered at this time.<br>Members can be studied in context and with the benefit of the presentation on clean and consistent drawings which will enhance comprehension of the overall structural function. |
| Loading assumptions | Before detail design is begun and after all modifications. This may imply a contradiction in practice. In this case it follows that the check must be repeated. |
| etc. | etc. |

The complete list will obviously be much longer but for most other subjects the answer to the question of timing is not difficult to find and to rationalize, without any scientific information beyond common sense.

How to Check?

Only an effective check is a check, everything else is make- believe and useless at best, counterproductive at worst, as it will create the illusion of quality, hiding its real absence.

To be effective, i.e. to catch serious errors with a useful degree of likelihood, a check must be:

- directed towards the essence of the subject,
- independent from circumstances which led to the original generation of the subject information and its errors.

Checking for the essential

For the engineer, the basic question about the essence of his work seems quite simple:

Will this thing which I am helping to create, perform as expected? This logically includes all the good properties of the structure:

durability
stability
safety
good appearance
etc.

and conversely it excludes all the possibilities of something being wrong, including all hidden faults.

As we have seen before, there is a right direction and a right time for the application of the filter. In order to be effective, the filter must also be of the right type. A filter with a mesh too fine will quickly become clogged, but a filter with too wide a mesh will let large - essential -particles slip through. The same will happen if the filter is too small to include the entire stream of the liquid, or if it is too large to enter the vessel in which the liquid is flowing.

In the reality of the concrete structure, this would translate about as follows:

Too fine a filter:

The person assigned to checking the placing of the steel will measure the position and diameter of every bar until he runs out of time - he will not be able to check the entire amount of information which is the right placing of the steel for, say, a slab, and whatever he was not able to see in his allotted time, will escape checking altogether.

Also the detail check of design calculations through repetition of the same is quite inefficient because the checker will likely be lured into the same traps as the author - the check is not independent as the filter was modelled after the original faulty method.

Too wide a mesh:

Someone who checks the slab reinforcing only in some spots, or only grosso modo, will not notice when the steel placer has cheated with the bars leaving out every tenth bar, to sell it later at his own profit. The structure is then going to be only 90% as strong as expected. This may not appear very serious, considering the usual safety margins, except when combined with some other essential ingredients, e.g. the faulty position of the bars which, due to the spotty checking, may also have escaped detection. Now the case may indeed become serious, the real safety margin becoming quite small. Add an occasional overloading, for instance during construction, and a failure may result.

Too small a filter:

This case would apply where someone checks the reinforcing steel who is not able to understand the structural function of the bars. It may then escape his detection that the stirrups in a beam are arranged in the wrong manner by mistake, for instance crowding them in the middle of the span and leaving them out near the supports. His filter did not include the filtering of this essential ingredient, and the beam in question may fail in shear - a particularly ugly mode of failure, because part of the information has escaped filtering.

Too large a filter:

Something similar will happen if the checking procedure includes questioning of too many things. Someone checking the slab reinforcing who worries about the formwork and the stability of the crane, as well as the concrete being poured into the column forms at the other end of the project, will not be able to spot misplacing of beam bars which were accidentally moved in longitudinal direction, leaving one end of a beam virtually unreinforced. This filter did not cover the stream of information it should have because it tried to include too many things at the same time.

Again, all this may appear trivial to the experienced builder, but it must be kept in mind that, due to restraints having to do with human or material resources, or with time, conditions are usually less than ideal, and circumstances may include strong incentives to create scenarios similar to the ones suggested above: Because no qualified personnel is available, the junior or apprentice is sent out. Or the senior engineer goes himself to check and sees a problem with the scaffolding, which draws his attention away from what he originally went out to see, which may have been the curing of the concrete.

In practice, perhaps not consciously but very really, one depends on other people attending to the construction, for instance the construction manager who will see something unusual, say the omission of some stirrups in a column, or the architect noticing that a large steel beam is fixed only by two bolts at one end, unlike the ten holding it up at the other end.

I am gladly confessing a multitude of cases where my attention was drawn onto faulty things by people whose concern had not really been to see them while they were attending to their own tasks. Needless to mention the importance of human relations in this context. Experience seems to indicate that jobs which are performed in adversary circumstances, or in a context of mistrust, are more error-prone.

## Who to Check

The obvious solution would seem to be: Send an experienced engineer, he is most likely to catch the greatest variety of serious errors (although he may let more minor things slip through). One more, the trivial answer, although quite appealing and in fact used very often, is not the real solution needed: Supermen are in short supply, and experienced engineers will tend to become overextended very quickly if they cannot learn to delegate work in a sensible manner.

How can we then, in a relevant way, answer the question about the choice of personnel when it comes to checking the slab reinforcing or the framing concept of a highrise building, or the computer analysis for a shell structure which is less than transparent, because it is so complicated.

From information theory we know that for a system, in order to be understood by a second system, that second system must be more complex than the first one. Free translation to our case yields about the following:

For the checker to be effective, i.e. to understand his subject, his knowledge must at least completely include the essence of the subject - he must master it. Thus, in the case of the reinforcing steel, if the purpose of the checking is simply the conformity with the drawings, then somebody without structural knowledge will do, if he can only count, measure and read drawings. If on the other hand, a last check of the design ought to be achieved at the same time, then the engineer must go and he must command at least enough knowledge about structures to be able to spot the effects of calculation errors - only ten bars where there should have been twenty, although the steel was placed in conformity with the drawing which already contained the mistake.

In the case of the framing concept, a interesting and disturbing conclusion can be drawn: If the concept came from the leading head of the engineering group, i.e., the one with the most and most complete knowledge, then logically nobody will be able to check the concept: The leader is all by himself with his decisions and where he does not master what he is doing, nobody else will. Also, nobody will consequently be able to spot errors in the concept. Sometimes, more or less consciously, the engineer in question will turn to such - seemingly superior - agents as the computer, in order to reassure himself. Doing so, he may but make things worse because the computer will not likely detect a conceptual error but faithfully repeat it in the form of modelling assumptions. In the mind of the engineer, however, the reassurance has been created through the impression that if it was possible to express the concept in a large quantity of five-digit numbers, it cannot be wrong.

In my career, I have been associated with a number of structures the complexity of which defied the real comprehension of even the best engineers, including in reality, if not in his own mind, that of the author of the concept. In all cases, the structure ended up being fed into the computer, using sophisticated and expensive methods and producing enormous amounts of numbers. Since no simplified way existed to make the concept accessible to somebody's judgement, this is where the matter ended, everyone per force having to accede to what the computer said. In reality, the matter, rather than having been checked, was made worse, since the work associated with the data processing certainly implied an important additional source of errors.

It must be recognized that the computer is, certainly not today, "a system of superior complexity" compared with the human mind. It can therefore not be trusted with a checking function when it comes to the complex and multidimensional situation which is a structural concept in all cases of unusual character, or of innovation.

Some say that the computer is an excellent learning tool. I cannot agree with this, even considering the advanced graphic tools of representation we have presently at our command. The computer will necessarily only teach back what it was given to "know" before, with a complexity somewhat reduced from the original version. If a wrong concept went in, what comes out will at best be equally wrong, if not worse.

When we therefore at our schools, free the students from the tedious work of analyzing structural situations by hand, in order to let them spend more time with the screen and type set, we are committing a capital mistake:

These people will find it very difficult to go beyond what the computer yields which is, after all, only the trivial function of turning numbers into other numbers, using mostly the very basic arithmetical operations. Already the effect of a decade of schooling with the emphasis on computer use begins to show. The minds of engineers fresh from school seem to be quite dev- oid of imagination, and it amounts to a Sisyphos-task to teach them the simplest elements of real design, e.g. a connection detail.

If this goes on, we shall have killed innovation soon in the field of structural engineering, and mistakes, as they will come out of the black box of the computer, will become increasingly difficult to catch. The computer is the ultimate fool, and we have elevated it to the ultimate authority. What is going to be the price?

## Some Tricks of the Trade

It was found that not everything can be checked, that the best time is not always an option, that circumstances are usually less than ideal and the right checker cannot always be expected to be available. What is it then that can be done to keep checking efficiency at an acceptable level, or even to improve it, in the face of this impossibility of the desirable conditions?

What counts is the end result, namely the real structure. Not the drawings and even less the calculations. Therefore: let us check the real structure or if this cannot be done, something as representative as possible: in the case of the wind load on the suspension bridge, perhaps a model with the necessary elastic properties. Or in the more mundane case like the slab reinforcing, instead of devoting all the allotted checking time on doing the design calculations over again, let us go out and look at the steel in place. The capacity of one particular piece of slab is quite simple and quick to establish on the spot, allowing for inelastic redistribution of moments, by the approximate expression

$$q = 8/l^2(M^+ + 1/2\ (M_1^- + M_2^-)$$

where

$q$ = the capacity (uniformly distributed load)

$M = A_s f_y \times 0.85d$

This expression will, within less than 10%, deliver the true capacity of the slab. Recognizing the approximate proportion of top steel to bottom steel will establish that no gross deviation exists from the usual in the less critical matter of steel distribution.

The checker can now quite easily proceed, looking at other portions of the slab, counting a few bars here and there, always comparing with the first piece where he made the basic evaluation.

This amounts to what I would like to call

Pattern checking

Pattern checking is essentially based on a kind of analogy theorem, about like this:

> "If a situation has been found satisfactory, then another situation which resembles the first one sufficiently can be assumed to be also satisfactory."

Therefore, if one slab section was found to be adequate by direct calibration, e.i. in this case a calculation of the capacity from physical evidence, then through visual comparison, other sections can be reviewed very quickly and efficiently, as long as they do not diverge from the first one in too many aspects. In the case they do, the observer will have to recalibrate his pattern, either by completely recalculating the capacity of another section, or by numerical extrapolation, e.g. in the case of a different span length, through multiplying the required steel section per width of slab by

$$\frac{l^2_{new}}{l^2_{old}}$$

Two things can be concluded from this example:

A calculation of the load bearing capacity, from physical evidence in the real structure, and subsequent pattern checking of similar situations, effectively filters all previous stations of information processing: The rightness or wrongness of calculations, of engineering and placing drawings, of bar lists and of the placing itself have become irrelevant because whatever errors may have been introduced, must become apparent in the final check which is directly considering the end result, rather than following intermediate steps - unless of course, the capacity calculation is wrong itself. However, the probability for the calculation to contain an error with precisely the same effect as the original design appears to be quite remote. In case of discrepancy, of course an error will be suspected and searched out.

The second and - with respect to checking efficiency -perhaps even more important conclusion is that pattern checking, if applied in a judicious manner, is an extremely powerful and efficient tool for error-detection. It is also a very old tool which is used by every experienced practician who, at a glance, will recognize something suspicious because it stands out - is noticeably different -from the familiar pattern.

The method is not limited to reinforcing steel, nor even to the checking of constructed reality. With the same approach, drawings or calculations can be reviewed very quickly. The method will turn up errors as "breaches" in the pattern if they exceed a certain threshold which, with some insight, can be set at a selected level, depending on the situation.

Patterns can be found for any aspect of a design or structure. They will of course be directed towards the target properties of the subject matter, e.g. the depth/span ratio of beams, whether on the drawing or at the construction site - if one wants to eliminate deflection problems. Or in the case of column load calculations, one out of a group will be calculated directly, and the rest of the group checked by comparison. In case considerable differences are found, a satisfactory explanation (different contributory area, more load from walls, etc.) either exists, or an error must be suspected which can now be searched for in a direct way, because it has been spotted.

Pattern checking is also a very efficient, and in many cases, the only possible way to deal with the verification of large quantities of information, such as - notoriously - those associated with computer calculations. It is important to note that pattern checking in order to be effective, relies very much on a favorable presentation of the data, especially in large quantity. Graphic representation has been found to be of the essence when checking computer output, if not the only viable way to deal with the enormous amounts of numbers produced in the context of even rather trivial tasks of analysis.

The experienced engineer will use pattern checking when looking at any type of data: He mentally compares the evidence before him with similar situations from his memory. This is probably the single most efficient error-filter which can be mobilized. Its only limitation consists of the things never done or seen before. In this case the checker will either be increasingly prudent as the novelty of the situation increases, or he will seek help from people for whom the situation is more familiar.

## References

Schneider, J., ed., Report of Rigi Workshop, IABSE, Report Vol. 47., Zurich 1983.

Melchers, et al., Human Error in Structural Reliability, Civil Engineering Reports 2/1984, 3/1984, 4/1984, 3/1985. Monash University.

# THE ROLE OF REGULATORY DOCUMENTS IN CONTROLLING H.E.

Heki Shibata*

Abstract

This paper deals with the role of regulatory documents in the prevention of human error in the processes of design through maintenance of critical facilities such as a nuclear power plant and a petro-chemical engineering plant. The author has tried to define several quantitative indices and to evaluate the regulatory capacity of such documents.

Uncertainty is treated as a usual randomness of parameters in Seismic PRA, however, those are quite different from each other, and uncertainty comes from "unknowness". The author discusses this problem, and ways to overcome it by a regulatory document.

## Introduction

The author has been working on the aseismic design development of critical facilities such as nuclear power plant and petro-chemical engineering plants. Each faculty stores materials which are extremely poisonous. If these escape outside of the plant area, a catastrophic disaster would come to the surrounding area. Since the configuration of equipment and piping systems in such plants is complicated, it is difficult to predict their behavior during destructive earthquakes.

As we have learned from accidents of nuclear power plants and petro-chemical plants throughout the world, most of the causes were operational errors, and some others came from fluid-structure interaction and deterioration. The author has been studying Seismic PRA, probabilistic risk assessment, as a part of this study, and has concluded that the effect of human error through design, construction, inspection, and maintenance processes was highly significant compared to other factors appearing in the reports of those accidents. This subject was discussed on various occasions: (a) 3-ICOSSAR'85 (Shibata, 1985a); (b) SMiRT-8 (Shibata, 1985b); (c) Workshop on Civil Engineering Application Fuzzy Sets (Shibata, 1985c). He is going to discuss it in the next two occasions: (a) Euro-China Seminar (Shibata,

*Professor, Institute of Industrial Science, University of Tokyo, 22-1, Roppongi 7, Minato, Tokyo 106, Japan.

1986a); (b) 8ECEE (Shibata, 1986b).

The author's survey indicates that human error had an extremely important role in the failure of various industrial facilities under past destructive earthquake conditions, (Shibata, 1985d) and that most of the causes of failures and accidents were human errors as mentioned above. This type of human error could be found only under a destructive seismic condition. In a case in Japan an ore-transporter was designed in 1941, and collapsed in 1968 during the Tokachi-oki earthquake. The details were discussed in one of the previous papers (Shibata, 1985a). A design error concerning one of the beams for seismic buckling led to the accident which killed one operator.

It is rather easy to see why failures occurred in such earthquakes, and how they relate to the human error problem, but it is very difficult to make its mechanism clear, and to find a means of prevention. Even though the author has been discussing this issue in several papers, he does not have a concrete answer.

Regulatory Documents

There are many kinds of documents describing the procedures of design, material selection, manufacturing, construc-tion, inspection and so on. In this paper, we call these documents as "Regulatory Documents". Standard Review Plan and Guidelines are included in this definition. In Section 6, the author will refer to "Code Capacity". This term denotes the ability to regulate a specified capacity of design objects, such as total strength of structures designed by the regulatory document.

If a regulatory document describes all information which is necessary to control the design of any object concerned, then the regulatory document is said to be "closed form" (Shibata, 1976). In this case, the descriptions throughout the document have no uncertainty, and every team is described clearly, except the fundamental knowledge related to the engineering matter of the object. This fundamental knowledge might be called "background knowledge". The author has pointed out (Tsutsumi, 1977) that the content of a regulatory document can be expressed by a regulatory tree, and this new type of tree is the twin of the fault tree, which has a top event to be controlled in the related regulatory document (Fig. 1(a) and (b)). The lowest events in the fault tree are the most concrete statements for the regulation. This is also true in the regulatory tree (in Fig. 2). Beyond these statements, we must expect basic engineering knowledge.

On the other hand, we eliminate some detailed statements from the document, for example, "the material should have enough strength and ductility, and be corrosion-free"

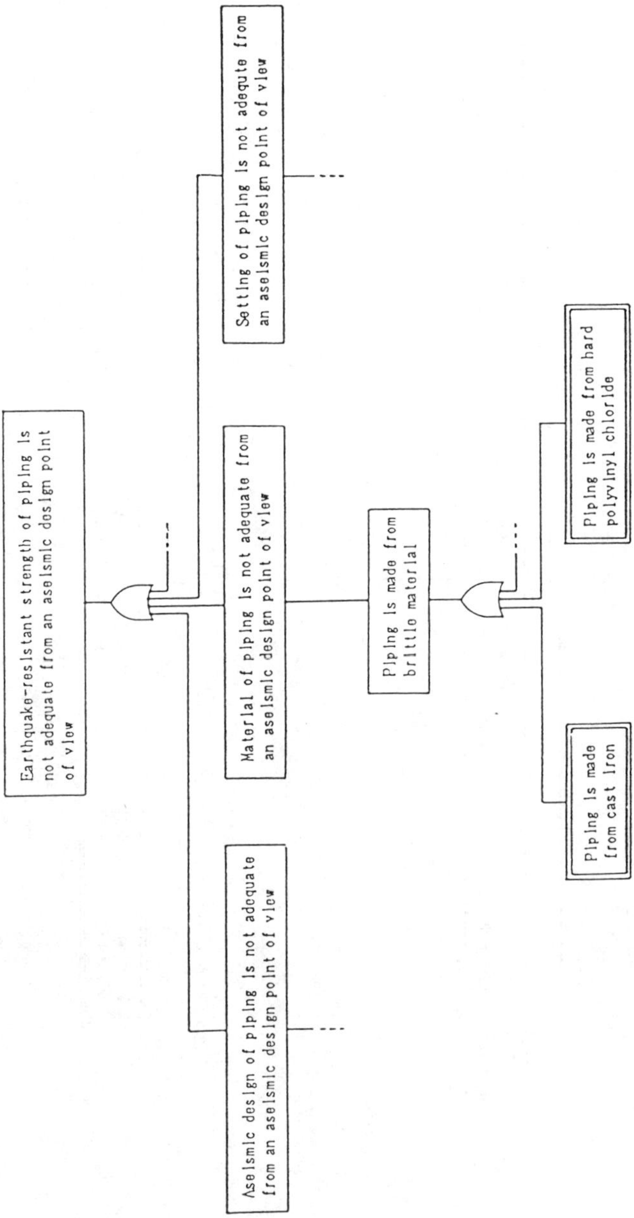


Fig. 1(a) Example of Fault Tree for Regulatory Document

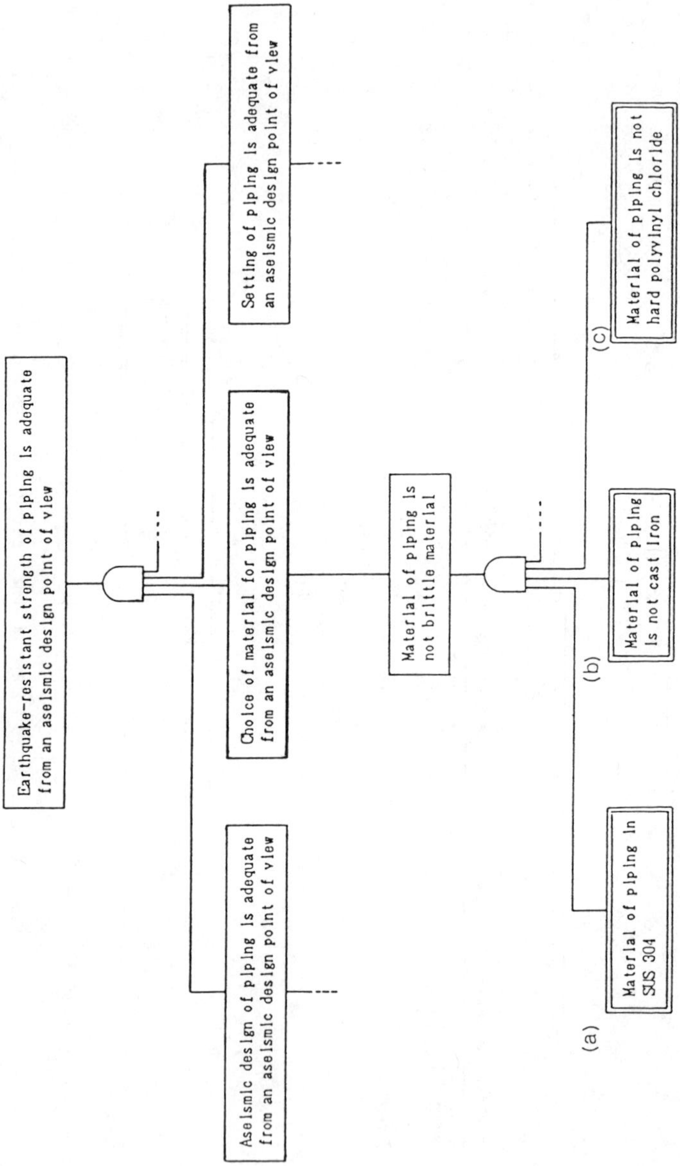


Fig. 1(b) Example of Regulatory Tree as a Twin of Fault Tree above

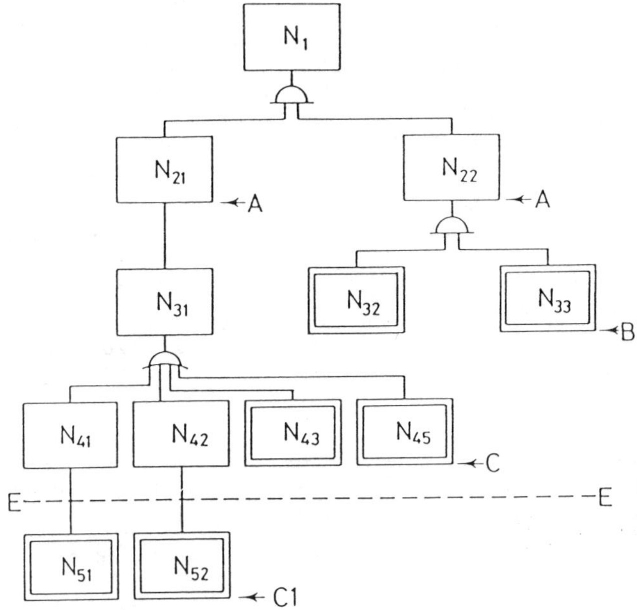


Fig. 2 Concept of Most-bottom Regulatory Statement

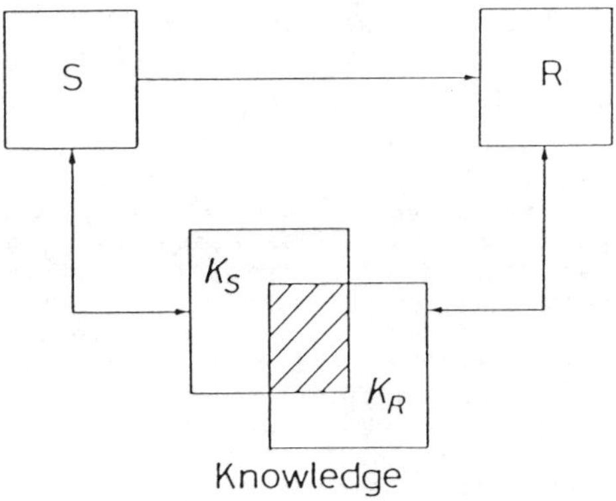


Fig. 3 Model of Communication Supported by Common Knowledge

instead of the statement "the material should be SUS 304". This means eliminating the bottom statements from the regulatory tree in Fig. 1. A regulatory document which lacks some statements may be called an "open form" as opposed to the previous "closed form". The tentative pressure vessel code of ASME in 1968 was a typical open form, however, the recent code of Section III is moving toward a more closed form.

Open form documents leave much freedom for engineers, but the form might cause design errors and other types of error through its process. An open form document, which gives more freedom to engineers, is the more ideal. However, it also gives a higher chance of error. It is difficult to eliminate such error without strictly described statements, that is, a closed form. This might be one use of a set of the statements describing design details.

Structure of Human Error

In the previous section, the simple term "error" was used. It means "human error". The author wants to consider four types of human error as follows:

i) simple error which is caused by a "wrong idea".

ii) communication error.

iii) Error which is caused by the "uncertainty" of the related phenomenon or the material used.

iv) error which is induced by a completely unknown fact.

If we consult a dictionary, we find two explanations of errors: "misunderstanding" and "wrong idea". A communication error is said to be a misunderstanding. The author discussed communication error at the previous NSF-Purdue Workshop (Shibata, 1985c). This type of error depends on the background knowledge of both sides, a sender and a receiver (see Fig.3). This type of error is more significant in such failures. The communication is supported by the common knowledge which is possessed by a sender and receiver. A sender, that is, a supervisor, and orderer, and documents like specifications, design guidelines, and so on, sends some amount of information to a receiver, that is, a design engineer, an inspector, and so on. But the amount of information necessary for actual performance is much more than the information which was sent by the communication. Therefore, there are many chances for errors. To improve such a situation, education to make the background knowledge uniform is important. Rearrangement, supplementation, interpretation and examples are necessary to eliminate communication errors. An illusion, the sender and receiver both having the confidence that they correspond with the correct common knowledge, is a cause of error. For this type of error, there is no way to use a computer to eliminate it, except by cross-checking large data files for each communication. However, this idea is nonsense, because requiring to check the common knowledge on each communication is almost impossible.

The modelling of this type of error is a random process also. Some sort of testing on communication may be used to establish the model.

The other two items will be discussed in the following section.

Uncertainty and Unknown

The author made the following comment at the Workshop mentioned above:

Comment on Uncertainty

Randomness: The random system can be clearly defined, therefore, we can estimate by observed data. If the sampled data has enough length, we can estimate the stochastic characteristic of they system with high accuracy. This means that we can estimate its governing equation with high accuracy also.

Subjective Probability: This probability can be improved, or reach more closer to its objective probability, if we can observe the phenomenon more often.

Fuzzy Characteristics: On the other hand, this is the information that is completely filtered, which cannot reproduce the original system or phenomenon. If it becomes reproducible, it is no longer "fuzzy information". And fuzzyness cannot be improved to more "crispyness" by frequent observations like subjective probability.

Vagueness/Ignorance: These are similar to each other. If an observer is ignorant of the system or phenomenon, the system seems to be vague. Fuzzyness is the expression of the observer's ignorance. The existence of a human is necessary for "Fuzzy Expression". The system expressed by fuzzy expression cannot establish its crispy governing equation as a random system, because it lacks the qualitative data to express it. Also "Chaos" is quite different. Its governing equation is clearly defined, but the state is chaos. If we can back trace its process to be chaos, we could know its governing equation. However, from this state, it is impossible to estimate it.

Again, Vagueness is only the result of the observer's ignorance, and the complexity in this discussion should be considered with the existence of the human, either observer or elements of the system. One person is complex enough to be a Fuzzy System.

In conclusion, the system would always be described by a set of definite equations. Its fuzzyness is only coming from the existence of the observer, unless it includes human or maybe some biological existence like a human in the system. Uncertainty in this sense is different from randomness, and it is not predictable either in deterministic or stochastic sense.

These definitions have not been established well. In the PRA study, the concepts of $\beta_U$ and $\beta_R$ are often used as shown in Fig. 4. As mentioned in the comment above, $\beta_R$ expresses the degree of randomness by the standard deviation of the stochastic variable which is random process, and it is clear in the stochastic sense. However, $\beta_U$ expresses the degree of uncertainty of the phenomenon itself. We cannot estimate the behavior of the variable even in the stochastic sense. We usually decide the value in a subjective way.

Here the author wants to discuss two examples. One example is the crash of an ore-transporter observed in Muroran during the Tokachi-oki earthquake in 1968. This crash was caused by the buckling of a particular element. That element was designed as shown in Fig. 5 in 1941. As a result of the error of the designer, one operator, who was in the carriage, was killed. If a design engineer had designed a structure without knowing a particular phenomenon, then it would result in failure. In this case, the buckling might be an unknown matter or uncertainty for him. The engineer who designed this ore-transporter might know the way of evaluating its buckling capacity, but he might not pay much attention to it. We estimate that the structure was designed for ordinary loadings excluding the seismic loading or with a very low-level seismic loading. As a result, the axial force on the beam might be only tensile force in his calculation. Now engineers are required to pay much attention to buckling of all elements and design "as Good design" in Fig. 5., whether their axial force is compressive or not. Although it was difficult to say that it happened through his error in 1941, we may say that it would be his design error in 1968.

In the above example, there is some ambiguity. The damage of oil storages in the Niigata-area by Nihonkai-chubu (Mid-Japan Sea) earthquake-1983, M=7.7, is another example. Even though the epicenter distance was approximately 230 km to the Niigata area, the single amplitude, wave-height, of sloshings of some storage tanks, whose eigen-periods are 9 ~ 10 sec, in this area exceeded 4 meters as shown in Fig. 6. This resonance-type response curve was obtained from the records, observed by scratches on their walls on various size storages in their area. The reason for such a high response is still discussed. There may be two main reasons. One is based on the local strata, and another one is the

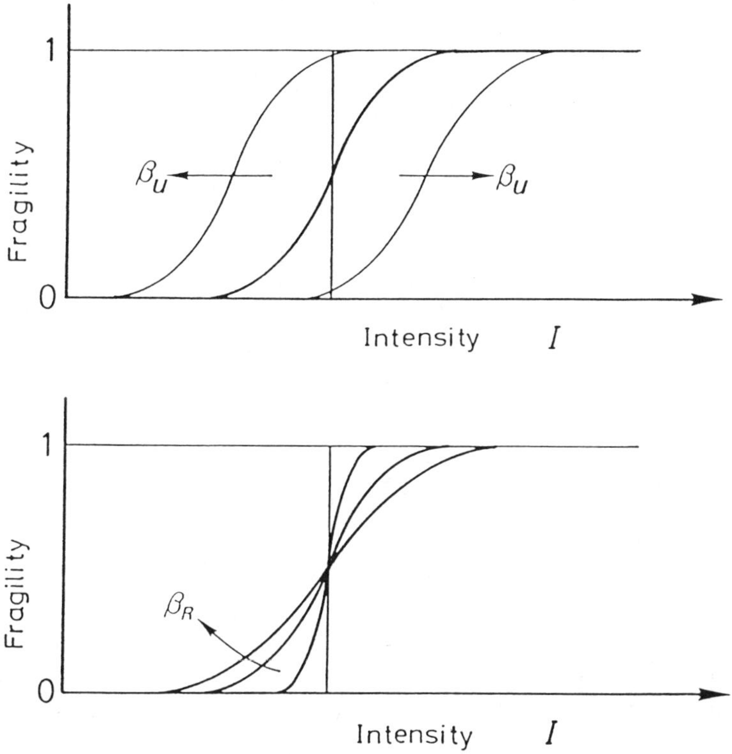


Fig. 4 Expression of Uncertainty $\beta_u$ and Randomness $\beta_R$ for Fragility Curve

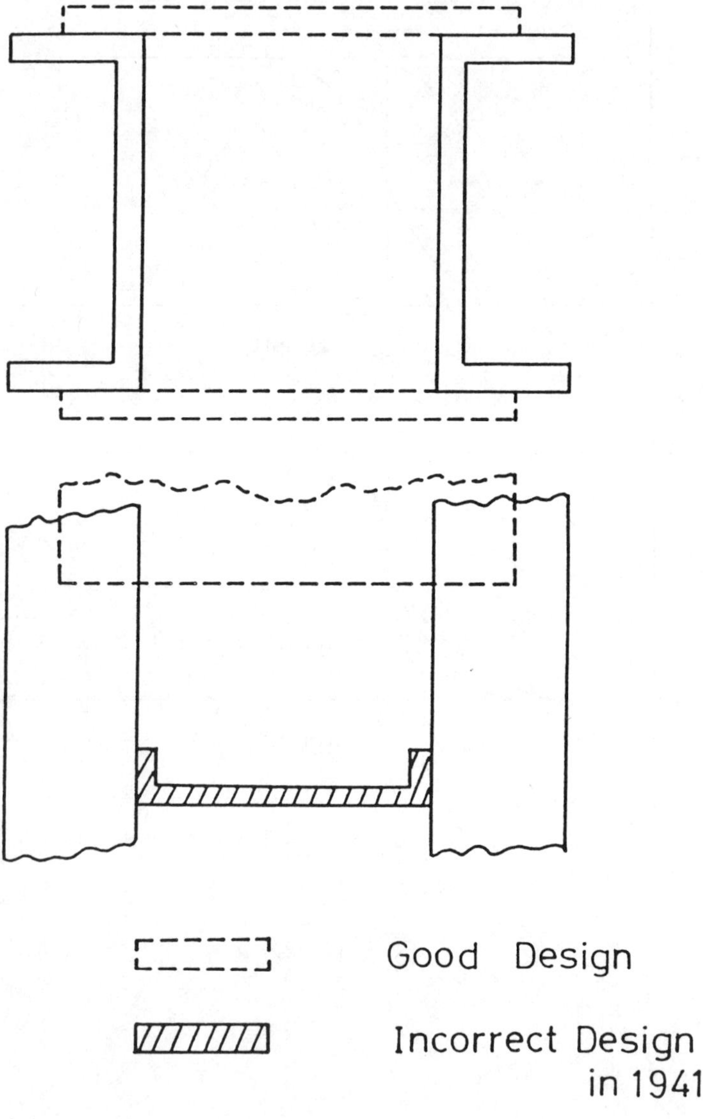


Fig. 5 Example of Human Error in Beam Design for Buckling Load, Ore-Transporter Crushed by Tokachi-oki Earthquake-1968

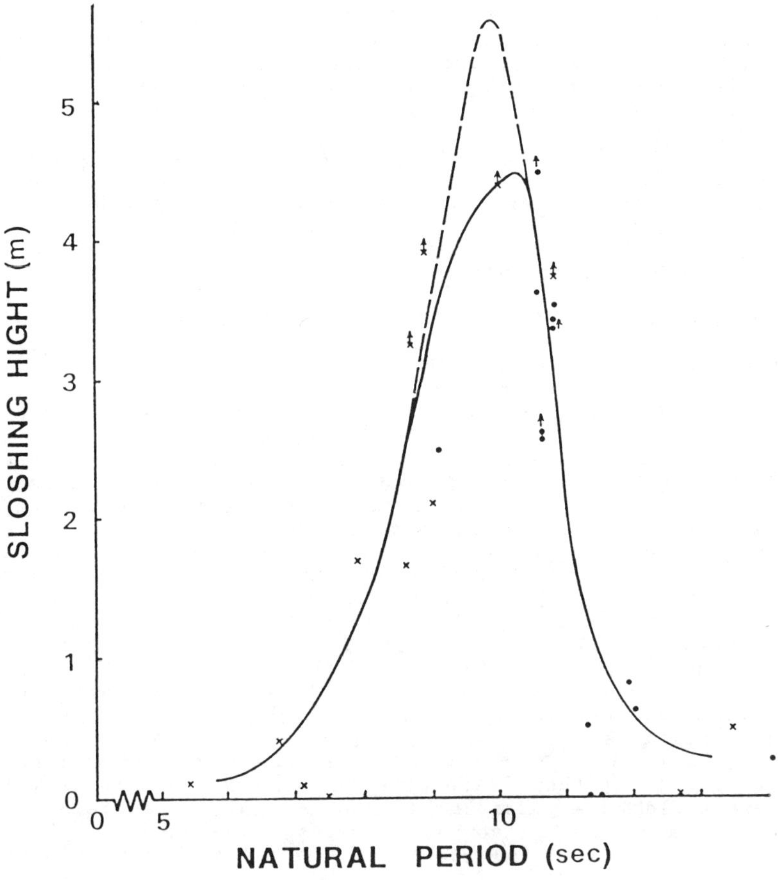


Fig. 6 Recorded Amplitude of Floating Roof of Oil Storages in Niigata-area by Nihonkai-chubu Earthquake-1983

failure process of the seismic fault according to seismologists. The former one should become engineering knowledge because we have experienced it. However, with the latter one, it is not clear, whether it was a spontaneous one or one that has a possibility of repeating a similar process in the future.

Our knowledge is always poor on natural phenomena including material behavior like deterioration, corrosion and so on, as well as ground motions, and response behavior. It is dangerous to treat such unknowness by $\beta_U$.

Computer Code and Regulation

The author referred to a communication error between a regulatory document and a design engineer in the previous paper (Shibata, 1985c). To avoid such an error, a computer code or program may help engineers. The author planned such a system for the Ministry of International Trade and Industry (MITI) (Shibata, 1981) in Japan on the aseismic design of high pressure facilities. The MITI order is an example of the regulatory standard, and it includes a static seismic coefficient method, a modified seismic coefficient method and a dynamic modal analysis method and other design criteria and details to design a reaction tower, a spherical tank, a cylindrical tank, supporting structures and similar facilities except piping systems. The content of this code may be substituted by the series of the computer program called a SEISMIT. The SEISMITs-prepared by MITI-SEISMIT-TW for reaction towers, SEISMIT-SP for spherical tanks, SEISMIT-HV for horizontal cylindrical tanks-are available now. If an applicant prepares input data from his drawings with the conditions of the siting and other items as designated in the standard, the result will be shown as a computer output, simply whether or not the design has any defects. If there would be any defects in the design, those items will be indicated in the computer output. The licenser checks the quality of related engineers, both for design and computer, in advance.

And on each design case, the licenser checks only the adequacy of input data, and the output result which designates "no defect". By using a standard format for input data, the human error for input process may be eliminated, and the check of them becomes easier. Human error in the interpretation of the regulatory standards and the use of equations will be prevented.

Since 1981, many engineers have been qualified to use this series of computer programs, and they have applied their results to the licensing of their structures.

Code Capacity

The author presented a concept "design ampleness" in

one of his previous papers (Shibata, 1979). The concept of a regulatory tree was also described in Section 2. "Open form" and "closed form" were discussed in relation to this concept. A closed form regulatory document has enough statements, and describes a matter completely. In such a case, the document has enough depth to regulate the matter. The "description depth" of the design procedure is one of three factors to measure a code capacity. The third one is "design detail", and will be discussed later.

The numerical evaluation of description depth or completeness is possible in principle. This was discussed in the draft (Shibata 1986) of the paper for the 8th European Conference on Earthquake Engineering, but this may not be presented at the conference. A detailed discussion will be made at a future occasion.

If the depth is insufficient, the chance of human error in the design process will increase, however, the freedom of design may give a chance for a designer to reach his heuristic design. This is really alternative.

## Design Details

Design detail descriptions are important to avoid human error in the design process. The author often refers to the utilization of general design criteria for the same purpose. There are some differences case by case. For example, the practice of putting hoop bars in the R.C. column, or the limitation of radius of a welded corner in the design detail, and the prohibition of the use of cast iron for flammable liquid lines or that of flange type connection for primary coolant loops in nuclear power plants. Both types of information are supplements to background knowledge for design engineers. But the former description is more related to the design and construction. Even though it is difficult to evaluate their effect quantitatively in some cases, we know that it is very effective in preventing accidents or failures under any conditions.

## Concluding Remarks

This evaluation of regulatory documents has found that it has a direct relation to the human error problem. Of course, we know the effectiveness of regulatory documents in preventing human error at the design stage as well as other stages by our experience. This study has attempted to evaluate it in a quantitative way, and some indices have been obtained. Evaluating real existing codes by this approach is still a long way off. At this moment, it is beneficial to improve the regulatory document only through conceptual means.

References

Ministry of International Trade and Industry, "Anti-Seismic Design of High Pressure Gas Facility," MITI Order No. 515 (Oct. 1981) (in Japanese), trans. by Shibata, H., "Draft of Anti-Earthquake Design Code for High-Pressure Gas Manufacturing Facilities," Report of ERS, No. III-5 (July 1981), 30 pp.

Shibata, H., "On Method of Evaluation of Anti-Earthquake Design Code of Industrial Facilities," Proc. of 7WCEE, Vol. 4 (Sept. 1979), p. 613.

Shibata, H., "Role of Human Error to the Safety of Critical Structure," Proc. of Int. Conf. on Structural Safety and Reliability, Vol. II (May 1985), p. 11.

Shibata, H., and others, "The Damage Assessment of equipment and Piping Systems under Seismic Condition Emphasized on their Failure Modes," Proc. of 8th Int. Conf. on structural Mech. in Reactor Tech., Vol. K(b), K17/1 (Aug. 1985), p. 273.

Shibata, H., "Effect on Human Error for Structural Design Under Seismic Loadings," Proc. of NSF Workshop on Civil Eng'g. Application of Fuzzy Sets, Purdue Univ., (Oct. 1985), p. 45.

Shibata, H., "Assumed Process of Piping Failure in Nuclear Power Plants Under Destructive Earthquake Conditions," Proc. of Euro-China Conf. on Earthq. Eng'g., Beijing (June 1986), under printing.

Shibata, H., "How to Design and Evaluate an Anti-Earthquake Code," Proc. of 8 Europ. Conf. on Earthq. Eng'g., Lisboa (Sept. 1986), under printing.

Shibata, H., "Damage to Piping Systems by Earthquake," Rept. on Piping Review Com., NUREG 1061, Vol. 2, Addendum (April 1985), p.C-1.

Shibata, H. "Some Comment on the Concept of Engineering Regulatory Code," J. of IIS, Inst. of Ind. Sci., Univ. of Tokyo, Vol. 28, No. 5 (May 1976), p. 235 (In Japanese).

Tsutsumi, T., and Shibata, H., "On Logical Treatment and Evaluation of Regulatory Statements," Bull. of ERS, Inst. of Ind. Sci., Univ. of Tokyo, No. 11 (Mar. 1977), p. 49.

# ERRORS IN CONSTRUCTION

Fabian C. Hadipriono* and Charles C.S. Lin**

## Abstract

Construction errors and human errors during construction produce enabling and triggering events that cause accidents (failures) in construction. In this paper, the occurrences of these errors are modelled as censored Poisson processes. The positive correlation among the errors requires the use of lower and upper bounds to determine their probability. A case study is presented to demonstrate the use of the models.

## Introduction

Many construction accidents are caused by errors that occur during construction; however, these errors have seldom been discussed. Only a few papers have addressed this topic. For example, Yamamoto and Ang (1985) discussed the effects of gross human errors on braced excavation. Lind classified human errors as those performed by the architect, engineer, contractor, and authority and developed a model to combine them (Lind 1979 and 1983). Hereafter, the terms "accident" and "failure" are used interchangeably, denoting the incapacity of a structure or its member(s) to perform as specified in the design and construction requirements. Only substantial failures leading to collapses are discussed here.

In this paper, errors during construction are classified as construction errors and human errors. A construction error (CE) is defined as an act performed intentionally or unintentionally in a construction operation, resulting in a construction performance that deviates from the acceptable practice outlined in the specifications. The term "acceptable practice" follows from Nowak and Carr's definition (1985): a technical practice that is acceptable to a significant number of the most knowledgeable authorities in the corresponding field. An example of a CE is the worker on a bridge construction project who intentionally neglects the use of washers when tightening A325 bolts with a calibrated wrench, believing that the washers are unnecessary despite the specification indicating their use. One may also

---

* Assistant Professor, Department of Civil Engineering, The Ohio State University, Columbus, OH

**Research Assistant, Department of Civil Engineering, The Ohio State University, Columbus, OH

unintentionally misset the automatic cutoff tightening of the calibrated wrench, resulting in the undertightening of the bolts. CE includes errors in judgment or actions that are directly related to the construction activities; such errors include poor workmanship, early falsework removal, or omission of bracing. A construction accident can also be caused by a non-construction error (CE'), which is an act performed intentionally or unintentionally outside a construction operation and results in a performance that deviates from the acceptable practice. An example is the inadequate size of the component of a permanent structure which fails during an acceptable timing of falsework removal. Here, the construction accident occurs not because of falsework removal but rather because of an error in component design. This paper, however, will emphasize CE.

In the following sections, the concept of CE is expanded and elaborated, probabilistic models are developed and their assessments are provided, and a case study is presented to demonstrate the use of the models.

## Errors in Construction

A CE may or may not be a human error (HE). Here, we define a construction related HE as a CE that originates from the person(s) who is (are) involved in a construction activity and is (are) capable of performing within the acceptable practice but fails to do so,. For example, an accident involving a crane which hits a falsework is an HE if the accident is caused by the operator's error in either judgment or in operation of the crane. If the error originates from the equipment or if the operator does not know how to operate the crane but is forced to do so and fails, we do not consider such errors as HE but rather an equipment error or procedural error, respectively. We consider these errors as non-human errors or HE'. Another example of an HE' is poor workmanship in falsework erection caused by a procedural error in hiring inexperienced workers. Hence, an HE' in construction is a CE other than HE. For the purpose of this paper, however, we emphasize the occurrences of HE in relation to construction only.

Usually, a CE includes enabling and triggering events (Hadipriono 1985; Hadipriono & Wang 1986). An enabling event (EE) is an event related to the resistance of falsework components to external events. An undesired EE usually occurs due to design or construction deficiencies (e.g., inadequate falsework bracing). A triggering event (TE) is usually an external event that precipitates component failure (e.g., lateral movement during concrete operations). Investigations show that an EE is almost always undesirable but that a TE may be normal or undesirable. In several cases, failure occurs when both an undesired EE and TE take place. In some cases, either one of these events is dominant and can cause construction failure. For example, if the falsework bracing is extremely deficient, falsework collapse is inevitable even under normal concrete operations. On the other hand, a worker may mishandle a derrick crane which hits a very stable falsework leading to a construction accident. Hereafter, we shall refer only to the undesired EE and TE that cause construction accidents.

The events described above are best represented by a Venn diagram as shown in Figure 1. The sample space includes both error (ER) and no error (ER'). ER is divided into CE and CE' or He and HE'. Emphasis is given to CE and HE only; hence, this event is further explored. Many types of EE and TE can be found in space CE, but our investigations are focused only on those that affect bridge falsework construction. The shaded area indicates the bridge accident (AE), caused by the EE, TE, or both in the domain of the HE for which we are interested in. The events described above are represented by the following probabilistic models.

Censored Poisson Process Models

In general, CE or HE occurs infrequently and randomly during construction activities as well as independently from one activity to another. Therefore, we can assume a Poisson process model for this type of error which is expressed by the following:

$$p_A(a) = \frac{(\lambda n)^a e^{-\lambda n}}{a!}, \quad a=0,1,2,\ldots; \qquad (1)$$

where A is the number of occurrences of CE or HE in construction activities; $\lambda$ is the mean rate of occurrences of CE or HE in a construction activity; and n is the number of activities in a project.

Note that from Figure 1, both events CE and HE intersect in an area indicated by CE∩HE, which shows that some part of HE is embedded in CE (this part is related to human errors in construction). Therefore, construction related event HE may or may not occur at each occurrence of event CE. We assume statistical independence of the occurrences of HE at consecutive occurrences of CE. Then the probability of the occurrence of event HE in CE can be represented by the following censored Poisson process:

$$p_B(b) = \frac{(\lambda_{CE}\, np)^b e^{-\lambda_{CE} np}}{b!}, \qquad b = 0,1,2,\ldots; \qquad (2)$$

where B is the number of occurrences of HE in CE, and p is the probability of the occurrence of HE in CE given CE, or P(HE/CE).

For the same reasons as above, the probability of events

EE and TE can be represented by censored Poisson processes as shown in Eqs. 3 and 4, respectively:

$$p_C(c) = \frac{(\lambda_{CE}npg)^c e^{-\lambda_{CE}npq}}{c!}, \qquad c = 0,1,2,\ldots; \tag{3}$$

$$p_D(d) = \frac{(\lambda_{CE}npg')^d e^{-\lambda_{CE}npq'}}{d!}, \qquad d = 0,1,2,\ldots; \tag{4}$$

where C is the number of occurrences of EE in construction related HE; D is the number of occurrences of TE in construction related HE; q is the probability of the occurrence of EE in HE given HE and CE, or $P(EE/HE \cap CE)$; and q' is the probability of the occurrence of TE in HE given HE and CE, or $P(TE/HE \cap CE)$.

Subsequently, assuming the independence of the occurrences of EE and TE, the probability of the occurrence of accidents due to the errors in construction, P(AE), shown as the shaded area in Figure 1, is found as:

$$P(AE) = \{1-p_C(0)\} + \{1-p_D(0)\} - \{1-p_C(0)\}\{1-p_D(0)\} \tag{5}$$

Clearly, in order to obtain the result of Eq. 5, information about $\lambda_{CE}$, p, q, and q' are needed. The following analysis is essential for this purpose.

## Probability Assessment

Previous evaluation on events CE, HE, EE, TE, and AE leads to the assumption that these events are usually positively correlated to a certain unknown degree. In other words, we expect that the increase of the occurrence of an event will increase the likelihood of the occurrence of the other; for example, the occurrence of EE will increase with the increase of the occurrence of CE. With this assumption, we can determine the lower and upper bounds of probabilities p, q, and q'.

First we determine the lower and upper bounds of $P(HE \cap CE)$ which are expressed as (Weed 1981):

$$P(HE)\,P(CE) \le P(HE \cap CE) \le MIN\{P(HE), P(CE)\} \tag{6}$$

If Eq. 6 is divided by P(CE), the result yields:

$$P(HE) \le p \le MIN\{\frac{P(HE)}{P(CE)}, 1\} \tag{7}$$

Note that in Eqs. 6 and 7, P(CE) is the probability of the

occurrences of CE in a project, and P(HE) is the probability of the occurrences of HE in a project. These probabilities can be found from Eq. 1 such that:

$$P(CE) = 1 - e^{-\lambda_{CE} n} \qquad (8)$$

and

$$P(HE) = 1 - e^{-\lambda_{HE} n} \qquad (9)$$

Similarly, the lower and upper bounds for $P(EE \cap HE \cap CE)$ are determined as follows:

$$P(EE)\, P(HE \cap CE) \leq P(EE \cap HE \cap CE) \leq MIN\{P(EE), P(HE \cap CE)\} \qquad (10)$$

Dividing Eq. 10 by $P(HE \cap CE)$ yields:

$$P(EE) \leq q \leq MIN\left\{\frac{P(EE)}{P(HE \cap CE)}, 1\right\} \qquad (11)$$

With a similar procedure, the lower and upper bounds for q' can be assessed as shown in Eq. 12:

$$P(TE) \leq q' \leq MIN\left\{\frac{P(TE)}{P(HE \cap CE)}, 1\right\} \qquad (12)$$

Note that $P(HE \cap CE)$ in Eqs. 11 and 12 can be solved through EQ. 6. Clearly, from the above, we can see that information regarding $\lambda_{CE}$, $\lambda_{HE}$, P(EE), and P(TE) are needed to obtain P(AE). The following case study illustrates the use of the above procedures.

## Case Study

For the purpose of this study, a sample bridge construction population in California is used. The bridges are concrete and consist of various numbers of spans. Several of these cases were reported in brief form by the contractors. Therefore, much of the following information was based on the authors' judgment, consultations with George Thompson of the California Department of Transportation, and previous assessments by other experts in human errors. Observations based on Thompson's data (Tompson 1981) reveal 43 bridge construction accidents in the a 23-year period (1957 till 1980), during which time approximately 6,000 bridges were constructed in California.

Evaluation of the accidents reveals 18 and 31 occurrences of EE and TE, respectively, indicating P(EE)=0.0030 and P(TE)=0.0052. Note that since we only account for the most significant event(s) contributing to the accident, only an EE, a TE, or both was considered in each case. Based on the above data, $\lambda_{HE}$ and $\lambda_{CE}$ are assumed to be 0.0002 and

0.0003, respectively (note that figures with the same magnitude are suggested by Swain and Guttman 1983)). These values indicate the mean rate occurrence of events in a construction operation/activity. An approximate average number of 200 contract activities was estimated for each bridge project. Eqs. 8 and 9 yield P(CE)=0.05824 and P(HE)=0.03921.

Substituting these values into Eq. 6 and 7 yields:

$$0.00228 \leq P(HE \cap CE) \leq 0.03921 \quad (13)$$

and

$$0.03921 \leq p \leq 0.67331 \quad (14)$$

The value of P(CE ∩ HE) in Eq. 13 is used to determine q and q'. The use of the lower bound will not only produce a very conservative range for q and q' but will also result in an incorrect assumption of independency between HE and CE, while in fact a high positive correlation usually occurs between CE and HE in construction. Therefore, for the purpose of this case study, the upper bound of Eq. 13 is chosen to determine the bounds of q and q' in Eqs. 11 and 12, resulting in:

$$0.00300 \leq q \leq 0.07651 \quad (15)$$

and

$$0.00520 \leq q' \leq 0.13262 \quad (16)$$

Subsequently, the upper bounds of Eqs. 14, 15, and 16 are used in Eqs. 3 and 4, which in turn, are used in Eq. 5 to obtain the upper bound of P(AE). Similarly, the lower bounds of Eqs. 14, 15, and 16 are used in Eqs. 3 and 4 to obtain the lower bound of P(AE) in Eq. 5. This results in the following:

$$0.00002 \leq P(AE) \leq 0.00841 \quad (17)$$

As expected, the result indicates wide bounds for P(AE). This finding is due to the compounding range when computing P(AE) based on p, q, and q'.

The above evaluation is based on the average 200 contract activities performed for the bridge construction in the case study. In general, the number of contract activities varies depending on the dollar amount of the projects. Therefore, the probability values for CE, HE, and AE are computed for various numbers of activities and are listed in Table 1. Note that as the number of activities increases, the range for the probability bounds decreases.

Table 1. Probability Value of Construction Accident for Various Numbers of Activities.

| | 100 | 200 | 300 | 500 | 1000 | 5000 |
|---|---|---|---|---|---|---|
| P(HE) | 0.01980 | 0.03921 | 0.05820 | 0.09516 | 0.18127 | 0.63212 |
| P(CE) | 0.02955 | 0.05824 | 0.08607 | 0.13929 | 0.25918 | 0.77687 |
| UP(AE) | 0.00829 | 0.00841 | 0.00854 | 0.00879 | 0.00945 | 0.01571 |
| LP(AE) | 0.00001 | 0.00002 | 0.00004 | 0.00012 | 0.00045 | 0.00774 |

## Summary and Conclusions

Several error events that frequently occur during construction were defined, evaluated, classified, and represented graphically for probabilistic modeling. Since the occurrences of a particular error event are embedded in other error events, censored Poisson process models were introduced to represent the number of occurrences of these events. The models involve the conditional events p, q, and q' representing the embedded error events which may or may not occur with the occurrence of the basic events. Because of the assumption of positive correlation among the error events, lower and upper bounds for evaluating the conditional events are required.

A case study of California bridge construction was presented to demonstrate the use of the models. An average of 200 activities per project was estimated based on the information provided to the authors. The result shows a rather broad range of P(AE) with an upper bound of 0.00841 as a result of the broad range of p, q, and q', upon which the computation of P(AE) is dependent. This figure is close to the probability of accidents due to EE or TE regardless of HE, or P(AC)=43/6,000=0.0072. The fact that the upper bound of P(AE) is greater than P(AC) accounts for the broad range of the conditional probabilities; this is also desirable, since P(AC) lies within the bounds of P(AE).

## References

Hadipriono, F.C., "Analysis of Events in Recent Structural Failures," Journal of Structural Engineering, Vol. 111, No. 7, July, 1985.

Hadipriono, F.C. and Wong, H.K., "Analysis of Causes of Falsework Failures," Journal of Construction Engineering and Management, Vol. 112, No. 1, March, 1986.

Lind, N.C., "Models of Human Error in Structural Reliability," Structural Safety, Vol. I (1983), pp. 167-175.

Lind, N.C., "Optimization, Cost-Benefit Analysis, Specifications," Proc. 3rd Int'l. Congr. Appl. Stat. and Prob. in Soil and Struct. Engrg., Unisearch, Sydeney, NSW, 1979, Vol. 3, pp. 373-384.

Nowak, A.S. and Carr, R.I., "Errors in Structural Models," 4th International Conference on Structural Safety and Reliability, ICOSSAR 1985, Vol. II, pp. 41-50.

Swain, A.D., and Guttman, H.E., "Handbook of Human Reliability Analysis with Emphasis on Nuclear Power Plant Application," NUREG/CR-1278, August, 1983.

Thompson, G.W., "Falsework Related Accident," The Office of Structure and Construction, California Department of Transportation, unpublished report, 1981.

Weed, R.M., "Analysis and Application of Correlated Compound Probabilities," Contractual Relationships: An Essential Ingredient of the Quality-Assurance System and other Quality-Control Papers, Transportation Research Record 792, 1981, pp. 41-44.

Yamamoto, M. and Ang, A. H.-S., "Significance of Gross Errors on Reliability of Structures," 4th International Conference on Structural Safety and Reliability, ICOSSAR 1985.

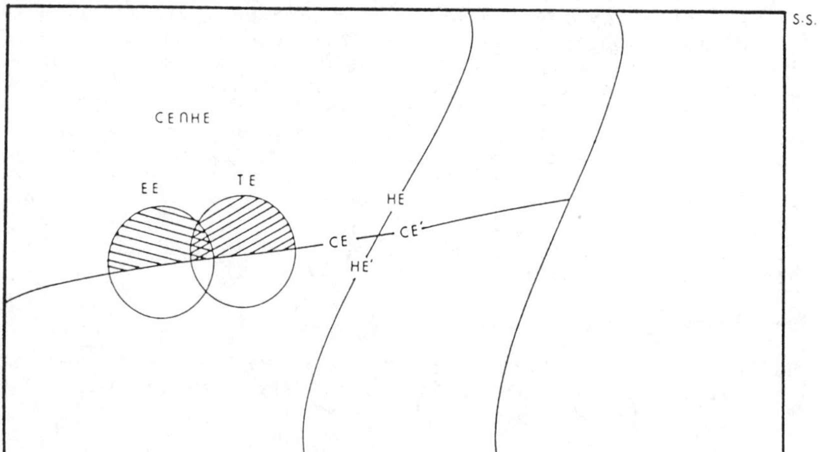


Fig. 1 Venn Diagram Indicating Errors in Construction.

# ON MODELLING ERRORS IN THE CONSTRUCTION INDUSTRY

Neal FitzSimons,* F.ASCE

Abstract

John Locke's classical discussion of human error is used to introduce a proposed model for the analysis of errors occurring in the construction process. Its characteristics are: simplicity, versatility, compatibility, practicality and feasibility. The model is based on the "triad", a systematic unit composed of input, throughput and output. Also discussed is the concept of the "teletic" system, which attempts to account for human characteristics analogous to the way in which a technical system accounts for physical characteristics.

## On Modelling Errors in the Construction Industry

In his "Concerning Human Understanding", John Locke, (1689), divided "science" (knowledge) into physica, practica and semiotica. The first dealing with "things"; the second with human actions and the third with "signs" i.e. communication. This trichotomy is directly applicable to the study of human errors in the construction industry. For example, when failures are investigated, three elements must be studied: the physical aspects (physica); the procedural aspects (practica) and the documentational aspects (semiotica).

In this same essay (book iv chapter xx), Locke also opines on four roots of human error. First, "want of proof"; accepting as fact that which should require proof. Second, "want of ability"; lack of experience, training or knowledge of the potential for error. Third "want of will"; not taking the time or trouble to prevent the error. Fourth "wrong measures of probability"; not gauging the situation properly from lack of perspective.

It is easy to relate each of these four roots to the construction industry. The engineer who accepts the information in a sales brochure as technical data may be guilty of the first "want". The architect who attempts a structural modification himself instead of calling upon a specialist may be guilty of the second. The draftsman who assumes someone else will check his work so he becomes careless may be guilty of the third. The purchasing agent who buys the cheapest products resulting in premature deteriora-

---

*Principal, Engineering Counsel, Kensington, Maryland

tion of the project may be guilty of the fourth "want".

For the construction industry it is both possible to and useful to construct a mathematical model to reduce human error impact which takes into account philosophical aspects such as Locke's thoughts while having other characteristics necessary for practical application.

The important characteristics of this model are:

1. Simplicity: simple to use and to understand by everyone involved.

2. Versatility: handles a variety of practical problems.

3. Compatibility: easily integrated into common models such as those used in construction planning.

4. Practicality: data required for input is available, economically attainable or reasonably assumable.

5. Feasibility: commonly available computer software and memory are sufficient.

One model that has these characteristics is based on "triads" (see Annex A). In it, each task is structured as a triad and all the triads are linked into a goal directed network just like the usual critical path network. Such a model can be as coarse or fine as necessary to put the problem into practical perspective.

Once the model is built, probabilities (or better still statistical data) can be associated with the performance level of each triad and its impact on the whole project can be evaluated. For example, the readiness, willingness and ability of the people associated with each task can be assessed from the standpoint of their propensity to err. What would be the effect of employing a "check list" for designers? What would be the effect of using an additional inspector (quantitative action) or a more experienced inspector (qualitative action) on reducing the impact of human errors on the project performance?

Any attempt to make an error reduction/evaluation model for a teletic system (see Annex B) such as that associated with a constructed project will be difficult. However, engineers with the help of colleagues in mathematics and the social sciences should be able to develop such a model which could become as common as CPM models. It would account for the total project; the planning, analysis, design, documentation, negotiation, construction, operation, modification, and ultimately demolition or dereliction of an engineering project.

When this model becomes available planners, designers, constructors, et al. will become more sensitized to the problem of human errors and its impact on project performance, and this, in itself, might well have a strong ameliorative effect. They will think more in terms of constructability, inspectability, maintainability, project organization and communication.

Further, the efficiency of effort and the chances of long-term success for a project might be enhanced by having an accepted procedure to determine which measures should be adapted to reduce the frequency, scope and severity of human errors in the total construction process. For example, with such a model, more realistic priorities could be made as could a better balance between the cost of an error mitigation/limitation measure and its value.

The practitioner would have a useful method of vulnerability assessment, but instead of the hazard being physical it would be psychological - the hazard of human error.

Annex A - A note on the Triad

The triad is to a system what the cell is to an organism; it is the basic functional unit. Just as most organisms consist of many differentiated cells so systems are composed of a variety of triads linked serially or in parallel. In modelling a system, only enough triads are used to serve the purpose of the model, from a few to hundreds.

The three components of the triad are organized so that the "throughput" acts upon the "input" to produce the "output". That is, output is input that has been changed in form, function, state, status or location.

In the construction industry, some triads are concerned with physical changes and others informational. An example of the first type would be the making of a concrete beam. In this case the input is cement, aggregate, water, reinforcing etc.; the throughput includes the mixer, the crane and, the workers; while the output is the completed, in-place beam.

An example of the second type is information transfer. The engineer ("throughput") operates on a variety of data sources ("input") to produce a set of plans and specifications ("output") for the contractor. In the next serially linked triad, the contractor ("throughput") using these documents and other data sources as input, operates on them to produce his "outputs", an estimate of cost for the project and his plan for executing it.

For practical purposes, each triad can have associated with it time, as in critical path analysis, cost, as in project estimating procedures or even probability of success, as in a plan for quality control. From breaking a project down into its basic triads, analysis; or building it up into a system description of the project synthesis, the user of the model based on triads can study the effects of change in any given triad on the performance of the whole system or can accumulate the effects of all triads to estimate its performance.

For project control purposes a network of triads can be constantly updated and better and better estimates can be made of final product performance or changes can be instituted if it appears that the expected performance levels are not met.

Annex B - Teletic Systems

A system is an orderly arrangement of units structured to facilitate the accomplishment of an objective. "Natural" systems are those which operate without human intervention in contrast to "Technical" systems which are artificially built to operate under well defined natural laws. "Teletic" systems are created by humans to achieve goals by the directed use of a society and its resources. Three teletic sub-systems are social, political and economic.

A social sub-system is concerned with the welfare of individuals, families or cultural groups.

A political sub-system is concerned with the setting of societal goals and encouraging their achievement; providing and enforcing societal rules; and the acquisition and disposition of the means for performing these activities.

An economic sub-system is concerned with the utilization of resources to meet societal demands for goods, works and services.

Normally, individuals or organizations (social, political or economic) have roles and interactions with all systems and sub-systems at a given time in a given place. Further, these roles and interactions are usually changing with time, place and circumstances.

As a simple example, most members of ASCE play a role and interact with a family and church groups (social subsystem), as voters and taxpayers (political sub-system) and as workers and consumers (economic sub-system). When they drive their car, they are using many technical systems and, unfortunately, polluting (to some degree) the natural system.

# DECISION MODEL FOR HUMAN ERROR IN STRUCTURES

Juan A. Melgarejo*, and Robert I. Carr**

## Abstract

An approach to model decisions in design and construction is presented. Decisions are based on structural reliability, sensitivity analysis, human error, cost of failure, cost of alternatives and cost of correction. The design and construction process has been modeled as a Markov chain. Although the complete model is still under development, we have learned that to make a good decision at any given stage of the process, a description of all possible alternatives for past and posterior stages is not necessary.

## Introduction

The objective is to develop a decision model that combines structural reliability (including human errors), cost of failure, and cost of alternatives so that decision-makers can rationally select the best alternative at any stage of design and construction. Without having to consider all possible alternatives at all other stages.

## Background and problem definition

Modern structural reliability analysis is based on a probability distribution for natural variations of loads and resistance. Current reliability model it based on error free cases. That is, they do not consider human errors and only take into account natural variation of load and resistance. The conventional models predict probabilities of failure which are ten to a hundred times smaller than the actual recorded rates. The reason for this discrepancy is believed to be human errors in the design-construction process. Rational decision making should be based on the understanding and consideration of natural variations, human errors, and cost-effectiveness.

Sensitivity functions and error frequencies offer a good conceptual approach to selecting control efforts. Combining sensitivity with error frequencies can identify the design and construction activities to which the structure's

---

*Research Assistant, Dept. of Civil Engineering, University of Michigan, Ann Arbor, MI 48109
**Professor of Civil Engineering, University of Michigan, Ann Arbor, MI 48109

and construction activities to which the structure's reliability is most sensitive. Consequently we should direct control efforts to those activities. However, decision making involves cost considerations as well, because design-construction activities require different amounts of resources to achieve the same reliability.

Though all of this is conceptually appealing, we do not yet have a decision model which allows us to focus our attention on specific activities or to select from among alternative design and construction processes. We are now developing a model which is designed to fill this need.

**Model's general description.**

We first define the context in which the decision is made. Figure 1 and 2 show a breakdown of the design-construction process, by such products as working drawings, shop drawings, materials, and installation. These products come from the different processes of design, detailing, fabrication, and construction. The processes are made of stages and substages, which are the different steps necessary to produce each product.

Design ----> Fabrication ----> Construction ---> Structure

- Process 1
    - Stage 1........Alternatives A1 and B1
- Process 2
    - Stage 1........Alternatives A2 and B2
    - Stage 2........Alternatives C2 and D2
- Process 3
    - Stage 1........Alternatives A3 and B3
    - Stage 2........Alternatives C3 and D3
    - Stage 3........Alternatives E3 and F3
        - Substage 1........Alternatives G3 and H3
        - Substage 2........Alternatives I3 and J3
        - Substage 3........Alternatives K3 and L3

The objective is to be able to make rational decisions among alternatives at one stage by characterizing other stages and processes without considering all possible alternatives at each of the stages and processes. The level of detail required in the characterization will be determined by the sensitivity of the decision to reduction of a more detailed break down of the process.

Two hypotheses are drawn from the past remarks:

1.- Decisions at one stage are sensitive to probabilities of errors and corrections at other stages and/or processes.

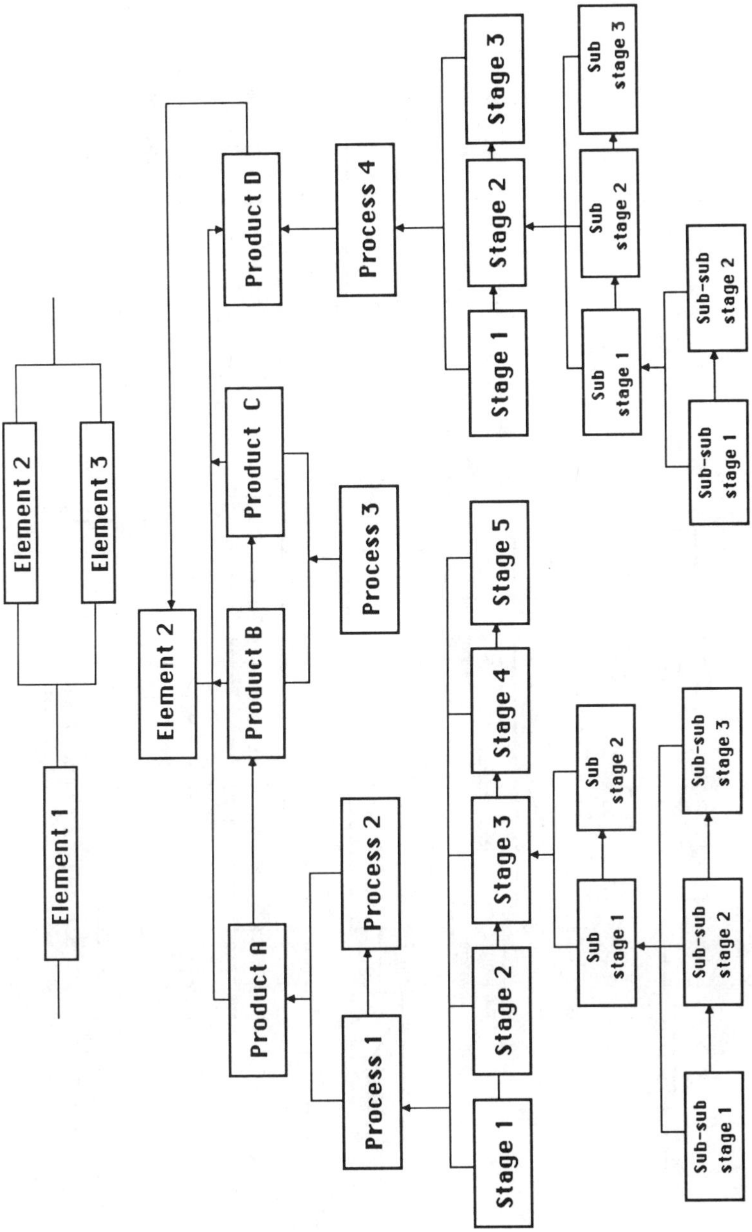
Element 1
Element 2
Element 3
Element 2
Product A
Product B
Product C
Product D
Process 1
Process 2
Process 3
Process 4
Stage 1
Stage 2
Stage 3
Stage 4
Stage 5
Sub stage 1
Sub stage 2
Sub-sub stage 1
Sub-sub stage 2
Sub-sub stage 3
Stage 1
Stage 2
Stage 3
Sub stage 1
Sub stage 2
Sub stage 3
Sub-sub stage 1
Sub-sub stage 2

Figure #1

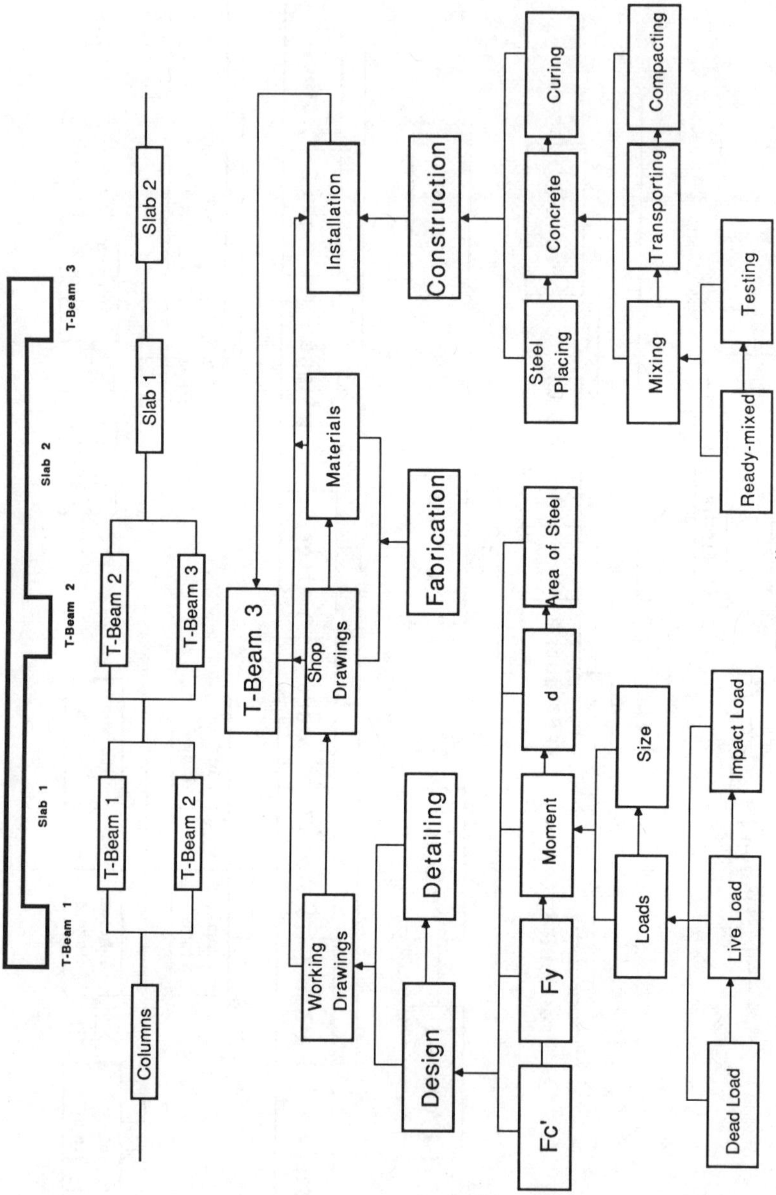
T-Beam 1
Slab 1
T-Beam 2
Slab 2
T-Beam 3
Columns
T-Beam 1
T-Beam 2
T-Beam 2
T-Beam 3
Slab 1
Slab 2
T-Beam 3
Working Drawings
Shop Drawings
Materials
Installation
Design
Detailing
Fabrication
Construction
Fc'
Fy
Moment
d
Area of Steel
Steel Placing
Concrete
Curing
Loads
Size
Mixing
Transporting
Compacting
Dead Load
Live Load
Impact Load
Ready-mixed
Testing

Figure #2

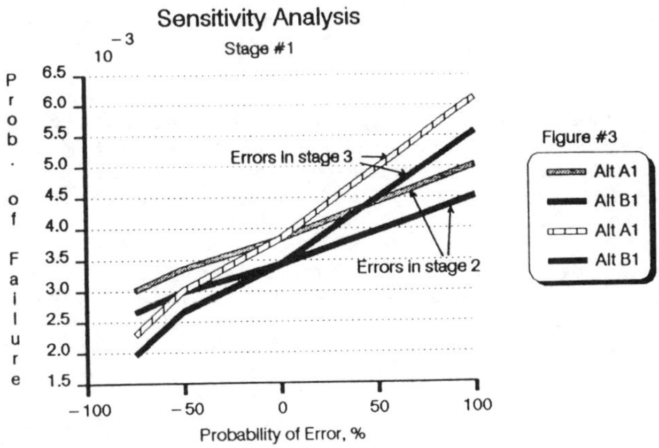


Figure #3

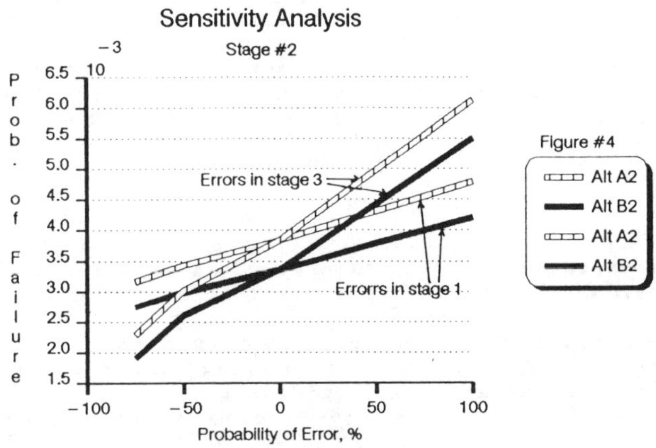


Figure #4

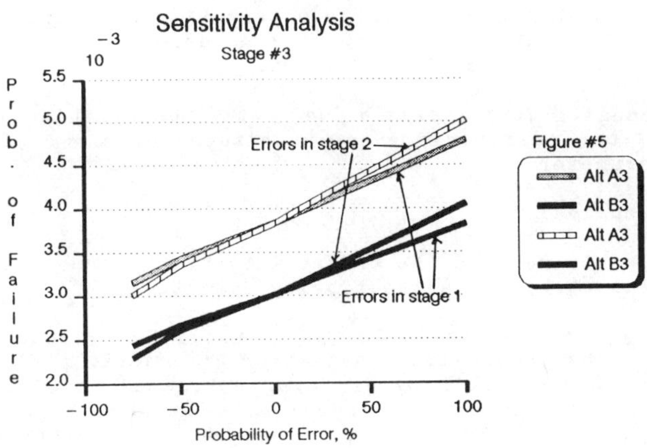


Figure #5

2.- These decisions are not so sensitive that we must consider all details of alternatives at other stages and processes.

We have taken a modular approach to characterizing the stages of the process. For example, when selecting between alternatives A1 and B1 in stage 1, we characterize all the alternatives in stage 2 as one module and characterize stage 3 as another module.

We hope to demonstrate that the model will handle decisions at the macro level as well as at the micro level. An example of a macro level decision may be the selection in the design stage between; (1) standardization of the girders or (2) use of a variety of girder sizes.

A decision at the micro level may be of the following type: within the construction stage, we identify the substage concrete placement, and within it we can identify these three sub-substages namely, concrete mixing, concrete transportation, and concrete compaction. For the sub-substage of concrete mixing we can choose between; (1) ordering ready-mixed concrete or (2) mixing concrete in the field.

Development

The alternative selection criteria should be based on the minimum expected cost of failure. This expected cost of failure considers the following variables:

a.) The probability of making errors [p(E)].
b.) The probability of correcting errors through inspection at any stage [p(C|E)].
c.) Sensitivity analysis for errors in design parameters.
d.) The estimated cost of the alternative selected.
e.) The estimated cost of correcting errors.
f.) The estimated cost of failure.

From sensitivity analysis we get probabilities that failure (f) occurs for different magnitudes of error (E) in a design parameter.

p(f|E remains)

Consequently we need the probability of the error reaching the final structure without being detected.

p(Er) = p(E remains in the structure). This depends on the probability that an error occurs, p(E), and its probability of being corrected, p(C|E), at the different stages and processes it goes through before the structure is completed and the error remains in the system (structure).

$$p(Er) = p(E) - \{p(C|E) * p(E)\} \quad (1)$$

Optimal design requires the minimization of the total cost, or the maximization of the structure's reliability at a particular total cost.

Therefore decisions among alternatives are made at each stage of the process based on the probability that an error arises in the stage considered and/or it is inherited from a previous stage and/or it is made in a future stage or process. Also included in the decision is the probability of finding those errors through inspection, the cost of repairing such errors, and the cost of the alternative selected.

Cost of failure in each failure mode is a function of the consequences of the failure. Those consequences may include property damage, human lives, personal liability, and damaged reputation. The cost of failure (Cf) is generally considered to be independent from the initial cost. The initial cost and the probability of failure are the two parameters which provide the trade-offs, after the type of structure and its use have been determined.

The cost of failure is the loss experienced by the liable participants derived from a structural failure. These participants may be individuals, private corporations, government agencies, etc. We must consider the cost to the person making the decision and costs to the entity or group affected by the decision.

At every stage (design, detailing, ..., construction) the decision maker will choose the alternative that minimizes his expected cost of failure.

$$E[C] = \{Cf * p(f)\} + Cc + Cs \quad (2)$$

Where:
- E[C] = expected cost of failure.
- p(f) = probability of failure
- Cf = cost of failure.
- Cc = cost of correction.
- Cs = cost of action or alternative chosen.

If we consider that errors occur in one parameter ($\alpha$) only, (from total probability theory) we produce:

$$p(f) = \sum_{X} p(f \mid Er_{\alpha} = X) * p(Er_{\alpha} = X) \quad (3)$$

Where:
- X = State of the error
- $p(f|Er_{\alpha} = X)$ = probability of failure given that the final state of the error in the design parameter is equal to X.

$p(Er_\alpha = X)$ = probability that the final state of the error in the design parameter is equal to X.

Errors may be corrected within one stage at a lesser cost than it would otherwise be if they are detected at a later stage when they may be impossible to remove and/or very costly to correct.

To have a reasonable assessment of what cost of correction might be for a given alternative at stage i, we multiply the cost of correcting an error from stage i at stage j (j different from i) times the probability of finding and correcting errors from i at stage j.

$$Cc = \sum_j Ccj \prod_{X>0} p(Cj|Ei = X)\ p(Ei = X)_{j-1} \qquad (4)$$

Where:

$Ccj$ = Cost of correcting in stage j an error originating in stage i.

$p(Cj|Ei=X)$ = Prob. of correcting an error in stage j given that the error in stage i is equal to X.

$p(Ei = X)_{j-1}$ = Probability that the state of the error which originated in stage i is equal to X after j-1 stages.

Substituting (3) and (4) in (2) we have the expected cost of failure, considering errors in one design parameter ($\alpha$) only, can be written as:

$$E[C] = Cf * [\sum_X p(f\ |Er_\alpha=X) * p(Er_\alpha=X)] + \sum_j Ccj \prod_{X>0} p(Cj|Ei=X)\ p(Ei=X)_{j-1} + Cs$$

To find the probability that the state of the error is equal to X ($Er_\alpha = X$) at the end of the process, we chose to model it as a Markov chain. Basically, a Markov chain is described by the conditional probability of arriving at a future state of the process, given the present state of the process, independently of how it arrived at the present state. These conditional probabilities are called transition probabilities, which can be described in a transition matrix.

As we know, all or almost all design parameters may have errors as they move along the different process stages. A stage may be composed of actions and/or corrections. An action can create errors but it cannot correct previous errors. On the other hand a correction may correct errors from selected preceding stages but it cannot create errors. Therefore, each stage may correct errors from other processes and/or from other stages. Correction has been considered as a function of the error state at time of correction.

To illustrate a Markov chain, we use the following simplified example that considers one product, one process and 3 stages; 1, 2, and 3. It considers only one design parameter which error state can take the value of 0, 1 or 2. The transition probabilities are read in the following way: The headings across the top of the matrix are the states of the errors as they arrived at the process, and the headings down the left are the possible states to which the state of the error may change after an action takes place. The values in the matrix are the probabilities by which the error state makes a transition from one state to another. Other nomenclature used in the example are described below:

x = random variable that represents the error magnitude at each of the stages.

$$X = \begin{cases} 0 & \text{for no errors.} \\ 1 & \text{for errors between 1 and 50\%} \\ 2 & \text{for an error larger than 50\%} \end{cases}$$

p(Ei=X) = The probability that the error in stage i is equal to X.

Ei = The matrix that defines the error probabilities of the action in stage i.

= [ p(Ei=0), p(Ei=1), p(Ei=2) ]

p(Cj|Ei=X) = Prob. of correcting an error in stage j given that the error in stage i is equal to X.

Cj|Ei = The matrix that defines the correction probabilities in the j stage given errors from stage i.

= [ p(Cj|Ei=0), p(Cj|Ei=1), p(Cj|Ei=2) ]

The number of Cj|Ei matrices at each j stage will be determined by the number of stage i errors that can be corrected at stage j.

The transition probability matrices would look like this:

$$
\begin{array}{cc}
 & \begin{array}{ccc} Er=0 & Er=1 & Er=2 \end{array} \\
Er=0 & \begin{bmatrix} p(E1=0), & p(E1=1), & p(E1=2) \end{bmatrix}
\end{array}
*
\begin{array}{cc}
 & \begin{array}{ccc} Er=0 & Er=1 & Er=2 \end{array} \\
\begin{array}{c} Er=0 \\ Er=1 \\ Er=2 \end{array} &
\begin{bmatrix}
p(C2\mid Er=0) & p(C2\mid Er=1) & p(C2\mid Er=2) \\
0 & 1-p(C2\mid Er=1) & 0 \\
0 & 0 & 1-p(C2\mid Er=2)
\end{bmatrix}
\end{array}
*
$$

$$
\begin{array}{cc}
 & \begin{array}{ccc} Er=0 & Er=1 & Er=2 \end{array} \\
\begin{array}{c} Er=0 \\ Er=1 \\ Er=2 \end{array} &
\begin{bmatrix}
p(E2=0) & 0 & 0 \\
p(E2=1) & p(E2=0) & 0 \\
p(E2=2) & p(E2>0) & 1
\end{bmatrix}
\end{array}
*
\begin{array}{cc}
 & \begin{array}{ccc} Er=0 & Er=1 & Er=2 \end{array} \\
\begin{array}{c} Er=0 \\ Er=1 \\ Er=2 \end{array} &
\begin{bmatrix}
p(C3\mid Er=0) & p(C3\mid Er=1) & p(C3\mid Er=2) \\
0 & 1-p(C3\mid Er=1) & 0 \\
0 & 0 & 1-p(C3\mid Er=2)
\end{bmatrix}
\end{array}
*
$$

$$
\begin{array}{cc}
 & \begin{array}{ccc} Er=0 & Er=1 & Er=2 \end{array} \\
\begin{array}{c} Er=0 \\ Er=1 \\ Er=2 \end{array} &
\begin{bmatrix}
p(E3=0) & 0 & 0 \\
p(E3=1) & p(E3=0) & 0 \\
p(E3=2) & p(E3>0) & 1
\end{bmatrix}
\end{array}
=
\begin{bmatrix} p(Er=0), & p(Er=1), & p(Er=2) \end{bmatrix}
$$

The underlying assumptions of the decision model are as follow:

- The probability of making an error is independent of the current state of the process. That is error probabilities among the stages are independent from one another. This is not an absolute restraint but it is useful for simplifying data requirements.

$$p(E2=X \mid E1=X) = p(E2=X)$$

$$p(E3=X \mid E1=X,\ E2=X) = p(E3=X) \quad \text{and so on.}$$

- The probability of correcting errors in one stage is independent of other corrections performed in preceding stages. It is also assumed that the probability of finding errors in one product is independent of finding errors in another product.

$$p(Ck \mid (C'j \mid Ei=x),\ Ei=x) = p(Ck \mid Ei=x)$$

$$p(Ck \mid Ei=x,\ Ej=x) = p(Ck \mid Ei=x) * p(Ck \mid Ej=x)$$

- The distribution that results from the Monte-Carlo simulation is assumed to be a normal distribution.

We have made an implicit assumption that hypothesis one is true because probabilities of correcting errors of one stage in another are important. However, we have not yet tested this to determine cases where it may not be true.

Hypothesis testing:

We performed a sensitivity analysis of the decision at one stage to different error frequencies at other stages to determine the accuracy of hypothesis two.

The probability of failure p(f) is plotted on the vertical axis and it was calculated as before:

$$p(f) = \sum_{X} p(f \mid Er_{\alpha}=X) * p(Er_{\alpha}=X)]$$

The sensitivity of the decision at one stage to changes in other stages, was determined in the following manner.

Changes in the error distribution were made one stage at a time. For example, to find how sensitive the decision in stage 1 is to changes in the error frequency distribution in stage 2, the error frequency distribution of stage 2 was varied for the different percentages shown in the graph. We held stage 3 constant at the error frequency given by the original distribution. Likewise, to find how sensitive the decision in stage 1 is to changes in stage 3, we varied stage 3 but stage 2 remained constant at its original error frequency distribution.

Figures 3, 4, and 5 show the sensitivity of probability of failure at one stage (vertical axis) to different probabilities of error at other stages (horizontal axis). The horizontal axis is the relative departure of the probability of error from an assumed input probability. As expected, probability of failure increases as the error rate increases. One also notices that the difference in slope between errors in stage 2 and stage 3 is due to a different number of corrections opportunities. In this case, stage 1 has 3 correction stages, stage 2 has two, and stage 3 has only one.

We can see that large variations in error rates in other stages did not change the alternative selected. We can conclude that there is a dominance by an alternative at a given stage even with large variations of error frequencies at other stages. This indicates that hypothesis two is true, though we may need to consider other stages and products to know the magnitude of probability of failure, we do not have to go into a detailed description of their alternatives.

Now, there are three other variables that make up the total expected cost (E[C]) on which decisions are based. They are the cost of alternatives (Cs), cost of failure (Cf) and cost of correction (Cc).

As we change anyone of these costs, the total expected cost also changes. Consequently it was interesting to notice the effects on the decisions as these costs changed.

As cost of failure increases from the original estimate ($40,000), which is represented by the 0%, the decision shifts towards a more reliable (less error-prone) alternative as long as savings in the expected cost of failure upset the cost increase, if there is any. Figure 6 represents such changes for stage 1. Similar results were found for stages 2 and 3 but are not presented here.

As we mentioned before, correcting errors once they leave our control may be very expensive and embarrassing. In Figure 7, the horizontal axis is the cost of such corrections, expresed as a function of the initial cost of the alternative (in this case Alternative A1). We can see that as the perceived cost of error correction increases, the model will favor a less error-prone alternative (alternative B1).

Conclusions

From these analyses we can conclude that in order to make a good decision at any given stage, a description of all possible alternatives for past and posterior stages is not necessary. However, we need the error transition probabilities of the stages as descriptors of the error state of the process.

[Preceeding error state of the process.] → [Transition probability matrix for the alternative being consider] → [Transition probability matrix for posterior stages.]

As cost of failure and/or correction costs increase, the decision model will favor the selection of a less error-prone activity. This will be so, up to the point where the incremental cost of this less prone activity is larger than the savings gained in expected cost of failure and expected cost of correction.

$$E[C] = \underbrace{Cf * p(f)}_{E[Cf]} + Cc + Cs$$

So, for alternative A,

$$E[C]A = E[Cf]A + CcA + CA$$

For alternative B,

$$E[C]B = E[Cf]B + CcB + CB$$

Considering that cost of failure is the same for both alternatives.

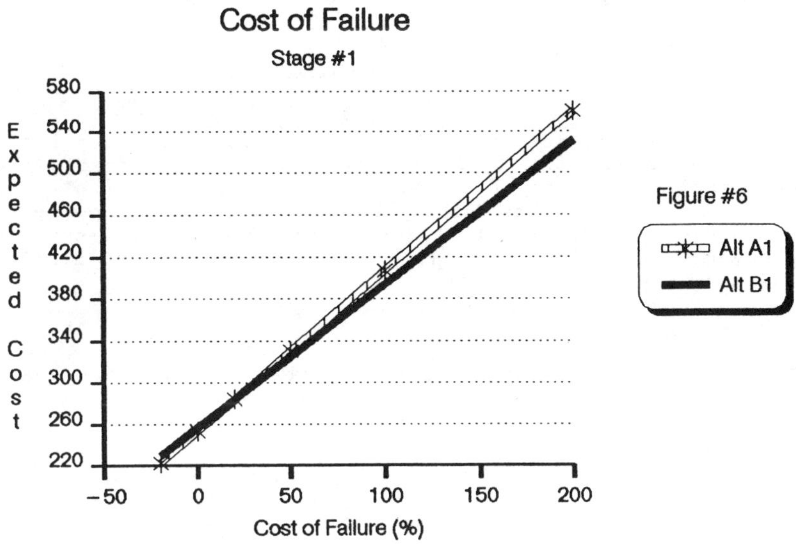


Figure #6

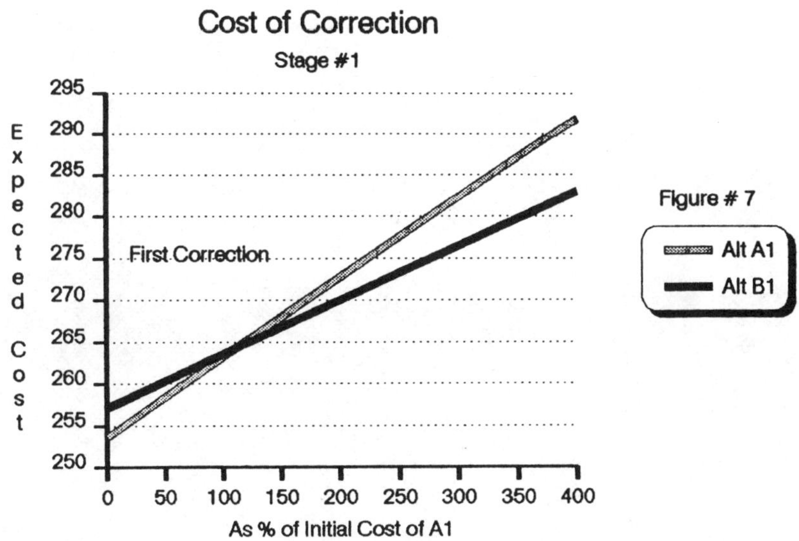


Figure # 7

Decision rule:

Select B if,

$$Cf * [ p(f)A - p(f)B ] + CcA - CcB \leq CcB - CcA$$

To make practical use of the model, it will require the error rates for the typical tasks used in the design and construction process. Therefore a data bank for such statistics is needed. The model is at the stage where real examples of design-construction processes are being considered, and we expect we will need to make some refinements and modifications.

## Acknowledgments:

This research was partially supported from the NSF grant CEE-8304962, which is gratefully acknowledged.

# PROBABILITY OF FRAME FAILURE RESULTING FROM HE

Norris Stubbs*

Abstract

A method for estimating the magnitude of the probability of failure due to human error in structures is proposed. The method consists of first (1) listing all possible scenarios of human error occurrences, (2) determining the frequency of occurrence of the scenarios, and (3) computing the probability of failure of the structure for the given scenario. Risk curves, in which the probability of frame failure due to human error is treated as the consequence, are constructed from the data provided in the first step. From the risk curves, the expectation of the probability of frame failure due to human error is computed. The latter value is taken as an estimate of the probability of frame failure due to human error. Numerical examples are provided to demonstrate the methodology.

## Introduction

In hurricane-prone regions, such as the Gulf and Atlantic coasts, emergency management officials are now considering the feasibility of utilizing deliberately selected multi-story structures to shelter inhabitants in the event of a hurricane. These structures, which are to have been fully engineered, are to be selected on the basis of the premise that they expose the sheltered evacuees to the least risk, when compared with all other competing mitigative alternatives. Designed and as-built plans as well as specifications are available for the potential structures. The data provided by these documents are to be supplemented with the results from a field survey of the building to provide current data for a risk analysis of the given structure.

In order to compare the option of using multi-story structures for shelter with other options (such as utilizing traditional shelters like school gymnasiums or churches, or physically departing the affected area and moving to higher ground) decision makers, who are generally not engineers, request that the risk of using a specific structure exposed to a given hurricane be expressed in easily "understandable" terms such as the expected number of fatalities of the occupants. One such method, to compute the expected number of

*Associate Professor, College of Architecture and Engineering, Texas A&M University, College Station, TX 77843

fatalities for a specific building subjected to a defined hurricane, has been developed utilizing Probabilistic Risk Analysis (PRA) techniques (Stubbs and Sikorsky, 1985). The technique requires as input, among other things, probabilities of failure for the structural frame, the foundation, the exterior cladding - including windows and doors, and the roof. However, if the end results are to be meaningful, estimates of the "true" probability of failure for each subsystem of the structure is required. Thus, for the existing structures, we need to estimate not only the "error-free" probability of failure but also the change in the error-free probabilities of failure due to human error and random hazards (Blockley, 1980; Yao, 1985).

Although techniques to evaluate the failure of structural systems are still evolving, the field is sufficiently developed to permit the estimation of the error-free probability of failure. Moreover, if local random hazards, such as flying debris, can be identified, estimates of the probability of failure can also be made for that category. However, in the case of the probability of failure of the structure resulting from uncovered human error, no general method has yet been devised to evaluate its contribution to the failure probability.

A method for estimating the magnitude of the change in the probability of failure of a frame from the occurrence of human error is proposed here. First, the technique of constructing arbitrary risk curves is reviewed. Next, using the risk curve conceptual model, the increase in the failure probability due to the existence of human error is assigned the role of the consequence. The presentation concludes with a demonstrative example.

## Risk Curves

The risk assessment process may be subdivided into a) the identification of hazard, b) the estimation of the risk given the hazards and the effected systems, and c) the evaluation of the risks (Burton and Pushchak, 1984). The evaluation of the risk phase usually involves a discussion of acceptable risks and requires extra-scientific input (Starr, 1969). The risk estimation phase essentially attempts to answer three questions: a) what can go wrong if the hazard strikes? b) how likely is it that the hazard will strike? and c) if it does happen what are the consequences?

A common approach used to represent risk is the risk curve (Kaplan and Garrick, 1981) or 'risk profile' (Fiksel and Rosenfield, 1982). The risk curve, denoted by R(x), is the complement of the probability distribution function of the annual losses and is defined to be:

$$R(x) = P[\text{annual losses exceed } x] \qquad (1)$$

where x is some realization of the random variable X. Closely associated with the risk curve above is the "conditional risk profile" which for a specific hazard E is defined to be:

$$R(x|E) = P[\text{losses exceed } x \text{ given that event } E \text{ occurred}] \quad (2)$$

Risk curves may be generated as follows. First a list of N scenario, each with its likelihood of occurrence and the consequence of the scenarios, is generated. Each line in the list is sometimes referred to as a triplet: e.g. the ith line will consist of triplet ($S_i$, $p_i$, $x_i$) - where $s_i$ is a scenario identification, $p_i$ is the probability of the scenario, and $x_i$ is the consequence of the scenario (Kaplan and Garrick 1981). Second, if the triplets are arranged as in Table 1 with the consequences ordered according to the rule $x_1 < x_2 < \ldots < x_n$, the cumulative probability, $P_i$, in the fourth column, can be determined by summing the $p_i$ terms from the bottom. Third, the family of points ($x_i$, $P_i$) are now plotted to give a staircase function. The smoothed curve through this stair case function is the risk curve R(x). A typical set of risk curves are shown in Fig. 1. Note that the expectation of the consequence E[X] may be given by:

$$E[X] = \int_{-\infty}^{\infty} x[d\{1-R(x)\}/dx]dx \quad (3)$$

which can be shown to be the area under the curve (Rowe, 1980).

Table 1. Construction of Risk Curves

| SCENARIO (1) | CONSEQUENCE (2) | OCCURRENCE (3) | CUMULATIVE PROBABILITY (4) |
|---|---|---|---|
| 1 | $X_1$ | $P_1$ | $P_1 = \sum_{k=1}^{N} P_k$ |
| 2 | $X_2$ | $P_2$ | $P_2 = \sum_{k=1}^{N} P_2$ |
| . | . | . | . |
| . | . | . | . |
| i | $X_i$ | $P_i$ | $P_i = \sum_{k=1}^{N} P_i$ |
| . | . | . | . |
| . | . | . | . |
| N | $X_N$ | $P_N$ | $P_N = \sum_{k=1}^{N} P_N$ |

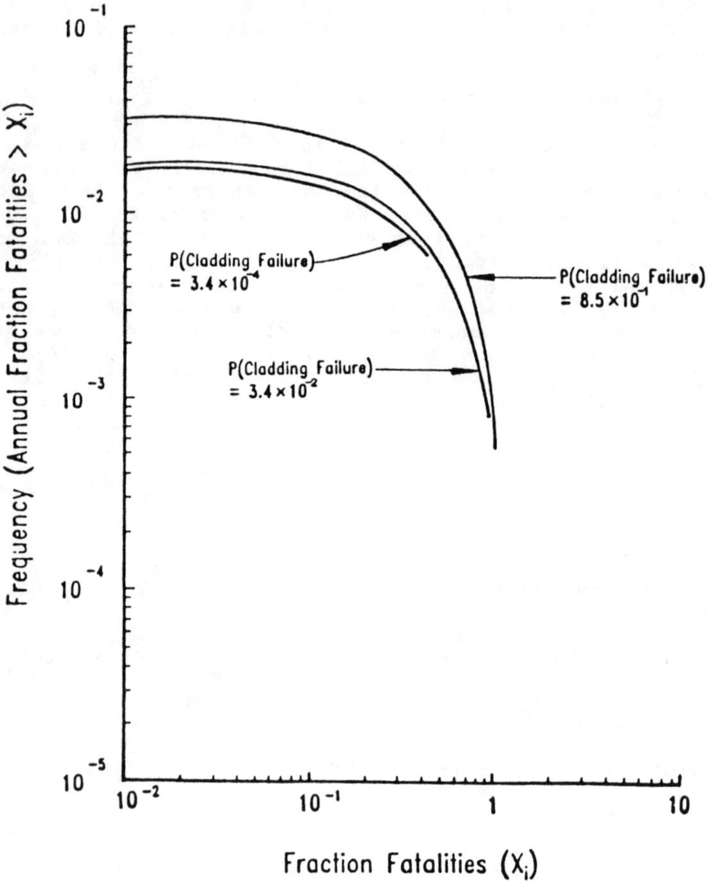


Figure 1. Typical Risk Curves

Risk Curves for the Human Error

Human error can occur during stages of design, fabrication, handling, construction or inspection. However, for the sake of illustration let us limit this discussion to unintentional human errors that occur during the construction process. In addition, to contain the complexity of the discussion, assume that the type of structure under consideration is a rigid frame. Assume also that the most likely construction errors will be the failure to correctly construct a rigid joint. Furthermore, if the constructed joint is not rigid then it is modelled as a pin joint. To be sure, in a realistic structure, more kinds of error may be considered; in addition, intermediate states of failure for the joint, for example modifying partial stiffness of a joint by a torsional spring, may be considered.

The first step in the risk curve generation process is to list the scenarios. In this case the first scenario may be described as follows: "As a result of human error during construction, the intended structural condition '1' (the rigid joint) has been erroneously replaced by structural condition 'm' (the pinned joint)." Since this error could have occurred at any joint, the consequence of the error is the increase in the failure probability of the frame by the amount $P_{e1}$ to reflect the errant pin joint at each structural node. The probability of occurrence of the scenario may be assigned some number $0 \leq q \leq 1$ which is related to the quality of the workmanship. Note that q will decrease as the quality of workmanship increses. For example, a basic equation to measure human reliability in the microelectronic industry (Dhillon, 1980), $R_n$, is given by:

$$R_n = 1 - e/n = 1 - q \qquad (4)$$

where n denotes the number of times the task was attempted and e denotes the number of errors. We may also assume that the rate of occurrence of the error follows, say, a Poisson's process

$P(k,\lambda) = \lambda^k e^{-\lambda}/k!$, $k = 0, 1, 2, \ldots$ The parameter $\lambda$ may be interpreted as a measure of the quality of workmanship.

Analogously, the second scenario becomes: "As a result of human error during construction, the intended structural condition '1' at exactly 2 joints has been replaced by structural condition 'm'". The increase in the probability of failure of the frame from the error-free state is now $P_{e2}$. This change, due to the introduction of these two new conditions, represents the consequence of the scenario. Assuming that the error rate, q, is constant and the error producing events are independent, the latter scenario has probability of occurrence $q^2$.

Similarly for the $i^{th}$ scenario we obtain the following

triplet:

$s_i$ = "intended structural condition '1' at i joints replaced by structural condition 'm'"

$x_i = p_{ei}$ = increase in probability of failure of the frame with i errors (5)

$p_i = q^i$

note that the i errors can occur in N!/i!(N-i)! ways.

Table 2. Risk Curve Data for a Typical Frame

| SCENARIO (1) | CONSEQUENCE (2) | OCCURRENCE (3) | CUMULATIVE PROBABILITY (4) |
|---|---|---|---|
| Error at 1 Joint | $P_{e1}$ | $q^1$ | $P_1 = \sum_{k=1}^{N} q^k$ |
| Error at 2 Joints | $P_{e2}$ | $q^2$ | $P_2 = \sum_{k=2}^{N} q^k$ |
| . | . | . | . |
| . | . | . | . |
| Error at i Joints | $P_{ei}$ | $q^i$ | $P_i = \sum_{k=i}^{N} q^k$ |
| . | . | . | . |
| . | . | . | . |
| Error at N Joints | $P_{eN}$ | $q^N$ | $P_N = q^N$ |

The results for a typical frame are listed in Table 2. Note that risk curves are given by the plots of Column 4 vs. Column 2. A series of Risk Curves for various values of q representing differing values of quality of workmanship is shown in Fig. 2. Finally, using the area under the curve to give the expected value of the consequences $E[P_e]$ - i.e., the expected increase in the probability of failure of the frame, given the scenarios of human error - is computed. A schematic representation of $E[P_e]$ versus q is shown in Fig. 3.

It is proposed here that the number $E[P_e]$ be taken as an estimate of the increase in the probability of failure due of the structure to human error. $E[P_e]$ will depend upon the loading environment, the error scenarios, the quality of workmanship, and the degree of inspection. Note that other statistics of $P_e$, for example $Var[P_e]$, may be invoked to

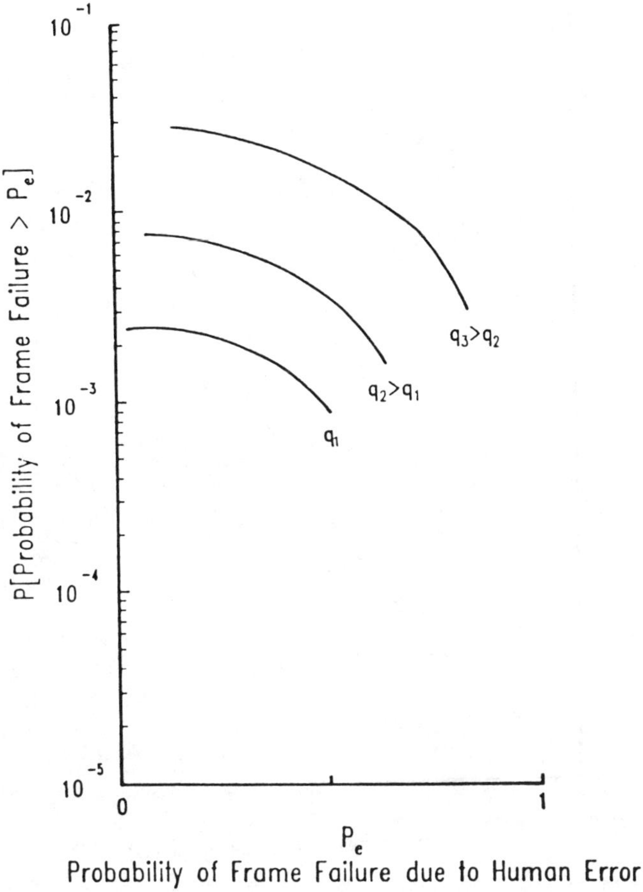


Figure 2. Effect of Human Error Magnitude on Risk Curves

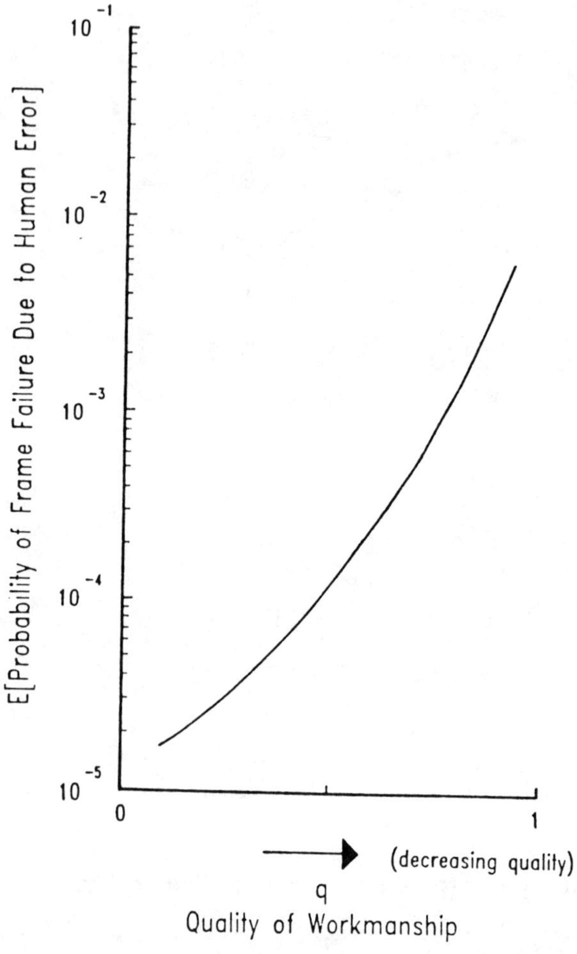


Figure 3. Schematic of Expected Value of Frame Failure Probability Due to Human Error Vs. the Quality of Workmanship

provide more confident measures of the probability of failure due to human error.

## Computation of the Change in Failure Probability Resulting From Human Error

Assume that we have two identical error-free structures, A and B, subjected to identical loading environments. Next, consider the following control group pretest - posttest experimental design:

$O_1 \rightarrow X \rightarrow O_2$

$O_3 \quad \rightarrow \quad O_4$

Where the symbols O and X represent, respectively, an observation and a change in the state of the structure (i.e., changes in the geometry or material properties). In structure A we make an initial observation $O_1$ then introduce some causal variable X (some quantifiable effect of human error) which is followed by a second observation $O_2$. The causal variable may be a modified joint or structural member. On structure B, i.e., the control experiment, we make an initial observation $O_3$ corresponding to $O_1$ and a final observation $O_4$ corresponding to $O_2$ but we introduce no change in the state of the structure. In the present experiment we may assume that observation $O_1 = O_3 = O_4$ and that it represents the probability of failure of the structure assuming no design, construction, or fabrication errors. On the other hand, $O_2$ is the observation associated with the probability of failure of the structure given the human error. Therefore, the present experimental design yields the symbolic causal relationship:

If X, then $(O_2 - O_1)$

Finally let N be the total number of joints in the structure and let $P_{f0}$ be the error-free failure probability of the as-designed structure subjected to some given loading environment. Then, if $P_{fij}$ is the failure probability of the frame with the $j^{th}$ arrangement of errors in the $i^{th}$ scenario, the increase in the failure probability of the frame due to human error (i.e., the consequence of the error in the language of the "Risk Curve" approach) is given by (ignoring higher order terms):

$$P_{ei} = \sum_{j=1}^{C_i} (P_{fij} - P_{f0}) \qquad (6)$$

where $C_i$ is the number of independent ways that i errors can occur with N joints. The value of $C_i$ depends strongly upon the number of possible errors (in this case N). The latter value, in turn, may be significantly reduced by means of inspection. If the nature of the human error is such that a negligible change in the state of the structure results, then $P_{ei} \simeq 0$. If the change in the state of the structure is significant $P_{fij} >> P_{fo}$.

A Demonstrative Example

To demonstrate the proposed methodology, the simple portal frame shown in Fig. 4A is evaluated for human error. Assume that the frame fails solely via a sideways mechanism and the random resisting plastic moment at each joint is $M_o$. Also, assume that significant errors can occur only at the joints. If an error has been made at a joint, the moment resisting capacity of that joint is set to zero. Considering only the case in which a horizontal loading is applied, the safety margin, Z, for the error-free structure is given by:

$$Z_o = 4M_o - Hh \tag{7}$$

where $M_o$ is the random plastic resisting moment for the columns and beam, and H is the magnitude of the random concentrated load. Therefore, assuming normally distributed variables,

$$P_{f0} = \Phi(-\beta) = \Phi(-\mu_z/\sigma_z) \tag{8}$$

where $\beta$ is the usual reliability index, $\mu_z$ and $\sigma_z$ are, respectively, the mean and standard deviation of Z and $\Phi$ is the standard normal distribution function.

If Scenario One, discussed above (the existence of one errant hinge) is in effect, the resulting safety margin for each of the four possible cases for the example structure is given by:

$$Z_1 = 3M_o - Hh \tag{9}$$

Therefore the probability of failure for each case is given by:

$$P_{f1j} = \Phi(-\beta_1) = \Phi(-\mu_{z1}/\sigma_{z1}) = P_1 \tag{10}$$

Using Fig. 4B the total probability of failure considering all four possibilities may be estimated by the formula:

$$P_{e1} \simeq \sum_{j=1}^{4} (P_{f1j} - P_{f0}) \tag{11}$$

$$= 4\,(P_1 - P_{f0})$$

Similarly, referring to Figs. 4C-4E it is readily seen that:

$$Z_2 = 2M_o - Hh,$$

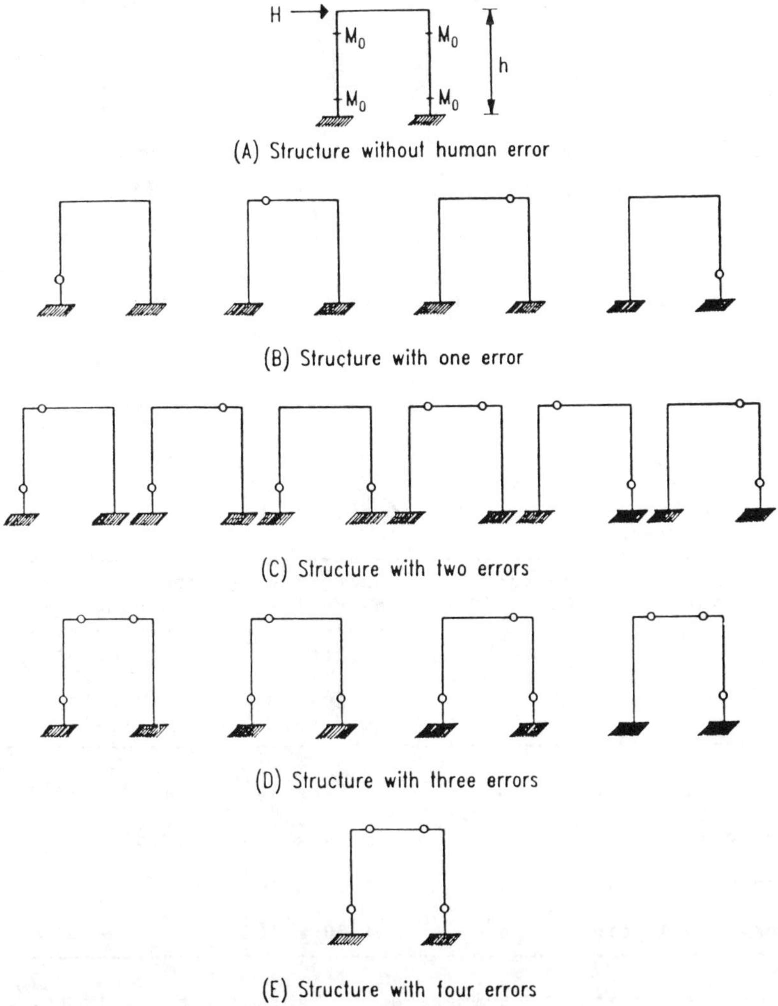


Figure 4. Example Structure with Error Scenarios for Errors at Four Joints

$Z_3 = M_o - Hh$, and (12)

$Z_4 = - Hh$

from which:

$P_{f2j} = \Phi(-\mu_{z2})=/\sigma_{z2}) = P_2 \quad j = 1,...,6$

$P_{f3j} = \Phi(-\mu_{z3})=/\sigma_{z3}) = P_3 \quad j = 1,...,4$ (13)

$P_{f4j} = 1 = P_4$

Substituting Eqs. (13) and (8) into Eq. (6) yields:

$P_{e2} = 6(P_2 - P_{f0})$

$P_{e3} = 4(P_3 - P_{f0})$ (14)

$P_{e4} \simeq 1$

Table 3. Frame Failure Dependence on the Number of Human Errors

| SCENARIO (1) | PROBABILITY OCCURRENCE (2) | CONSEQUENCE (Probability of Frame Failure Due to Human Error: Error at 4 joints (3) | Error at 2 joints (4) |
|---|---|---|---|
| Error at 1 Joint | q | $3.52 \times 10^{-6}$ | $1.76 \times 10^{-6}$ |
| Error at 2 Joints | $q^2$ | $2.34 \times 10^{-2}$ | $3.90 \times 10^{-6}$ |
| Error at 3 Joints | $q^3$ | $1.00 \times 10^0$ | --- |
| Error at 4 Joints | $q^4$ | $1.00 \times 10^0$ | --- |

These results are summarized in Table 3 for h = 12 ft, H = $12^k$, $M_o$ = 120k-ft., Cov[M] = 0.1, and Cov[H] = 0.2. Note that the results for errors occurring at only two joints are also presented (Fig. 5). This latter condition may result from an inspection program which reduced the uncertainty in the number of possible error locations. Risk curves generated using the results in Table 3 for various frequencies of human error (q = 0.1, q = 0.01, and q = 0.001) are shown in Fig. 6. The expected values of the consequence of each curve shown in the latter figure, as a function of the frequency of human error, are listed in Table 4.

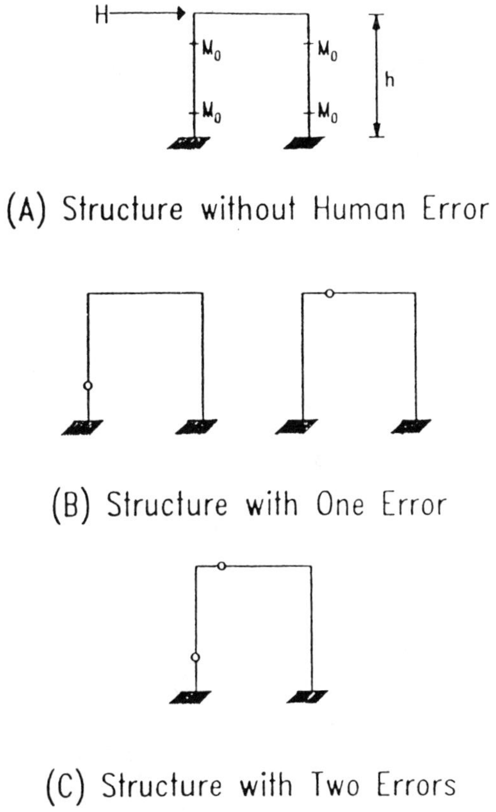


**Figure 5. Example Structure with Error Scenarios for Errors at Two Joints**

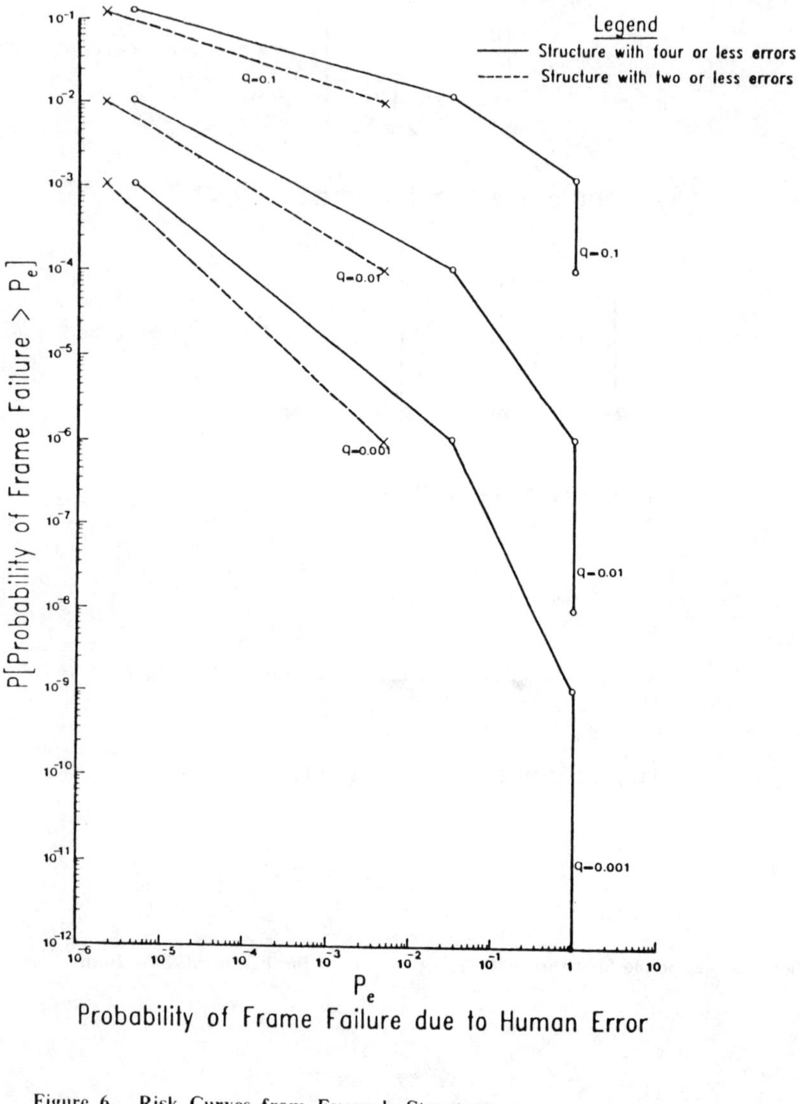


Figure 6. Risk Curves from Example Structures

Table 4. Expectation of the Probability of Frame Failure as a Function of Frequency of Human Error and Maximum Number of Error Locations

| FREQUENCE OF HUMAN ERROR | EXPECTED PROBABILITY OF FRAME FAILURE ¦ HUMAN ERROR | |
|---|---|---|
| | 4 locations | 2 locations |
| (1) | (2) | (3) |
| 0.1 | $1.33 \times 10^{-3}$ | $3.92 \times 10^{-5}$ |
| 0.01 | $3.39 \times 10^{-6}$ | $4.08 \times 10^{-7}$ |
| 0.001 | $2.79 \times 10^{-8}$ | $5.56 \times 10^{-9}$ |

These results show that (1) as the frequency of human error decreases, so does the expected probability of failure of the frame due to human error; (2) if the number of locations of possible human errors can be reduced by whatever means, (e.g., via inspection), the expected probability of frame failure correspondingly decreases; (3) for higher frequencies of human error, the probability of failure due to human error may be several orders of magnitude greater than the error-free probability of failure. These results are consistent with observations of, and opinions held by, other investigators (Nowak, 1985; CIRIA, 1977; Knoll, 1982; Blockley, 1980).

Summary and Conclusions

A method for estimating the magnitude of the in creasein the probability of failure of a structure due to human error has been proposed. The method consists of first (1) listing all possible scenarios of human error occurrences, (2) determining the frequency of occurrence of the scenarios, (3) computing the increase in the probability of failure from the error-free state for the given scenario. Second, risk curves in which the probability of frame failure due to human error is treated as the consequence in the risk context are constructed from the data provided in the first step. From the risk curves, the expectation of the probability of frame failure due to human error is computed. The latter value is taken as an estimate of the probability of frame failure due to human error. A numerical example was provided to demonstrate the methodology.

Acknowledgements

The financial support of the National Science Foundation under Grant CE 83-09511 is gratefully acknowledged. The assistance of Patricia Lombard in preparing this manuscript is also acknowledged.

References

Blockley, D. I., "The Nature of Structural Design and Safety," Ellis Horwood Ltd., Chichester, England, 1980.

Burton, I. and Pushchak, R., "The Status and Prospects of Risk Assessment," Geoforum, Vol. 15, No. 3, 1984, pp. 463-475.

Dhillon, B. S., "Reliability Engineering in Systems Design and Operation," Van Nostrand Reinhold Co., New York, 1983.

Fiksel, J. and Rosenfield, D. B., "Probabilistic Models for Risk Assessment," Risk Analysis, Vol. 2, No. 1, March 1982, pp. 1-8.

Kaplan, S. and Garrick, B. J., "On the Quantitative Definition of Risk," Risk Analysis, Vol. 1, No. 1, March 1981, pp. 11-27.

Knoll, F., "Human Error in the Building Process-A Research Proposal," IABSE Journal, Vol. 17, No. 4, 1982, pp. 35-50.

Nowak, A. S. and Carr, R. I., "Classification of Human Errors," Structural Safety Studies, Proceedings of the Symposium, 1-2 May, 1985, Denver, Colorado, Eds. J.T.P. Yao, R. Corotis, C.B. Brown, and F. Moses, ASCE, New York, 1985, pp. 1-10.

Rowe, W. D., "An Anatomy of Risk," John Wiley & Sons, New York, 1977.

Starr, C., "Social Benefit Versus Technological Risk," Science, Vol. 165, September 19, 1969, pp. 1232-1238.

Stubbs, N. and Sikorsky, C., "Fault Tree Analysis as an Integrative Technique for Safety Assessment in a Wind Hazard," Proceedings of the Fifth U.S. National Conference on Wind Engineering, November 6-8, 1985, Texas Tech. University, Lubbock, Texas, pp. 2B-1-2B-8.

______, "Rationalization of Safety and Serviceability Factors in Structural Codes," CIRIA Report 63, Construction Industry Research and Information Association, London, July 1977.

Yao, J.T.P., "Safety and Reliability of Existing Structures," Pitman Advanced Publishing Program, Boston, 1985.

# AN A.I. TOOL IN THE STRUCTURAL SAFETY CONTROL

D.I. Blockley*

Abstract

The research is aimed at analysing case studies of failures and successes in structural engineering and using the results to develop a computer based management tool which might provide some early warning of likely failure. The work is based upon a social science methodology involving grounded theory to develop a knowledge base which can be accessed through an appropriate A.I. language such as FRIL (Fuzzy Relational Inference Language).

Introduction

"Engineers and technologists in their quest to discover ways of organising nature and flushed with their successes in the physical sciences have perhaps neglected their reliance on human infallibility. In engineering only the product, the hardware, is a physical system; the system which designs it, produces it and uses it, is human and, therefore, complex and vulnerable", (Blockley 1980).

"It is gradually becoming clear that many disasters and large scale accidents display similar features and characteristics, so that the possibility of gaining a greater understanding of these disturbing events is presented to us ... (engineers can) ... build better, safer means of control and improve communications about hazards and safety procedures. What is now needed is the rephrasing of these questions in a sociological and psychological form so that research questions may be asked not just about the technical expertise and practices which surround the prediction and regulation of energy discharges (failures) but about the social context within which such practices develop and are carried out" (Turner 1978).

My work on human error at Bristol was, by 1980, reaching an impasse; how far should the engineer go in trying to study the problem as a social scientist might? At the same time Turner(1978), a sociologist at the University of Exeter, UK, had identified overall features of industrial failures which he could describe linguistically. His impasse was; how could he organize the patterns that he recognized into a directly useful form? It was natural,

---

* Reader, Department of Civil Engineering, University of Bristol , Queen's Building, University Walk BS8 1TR, U.K.

therefore, that we should work together: our skills and motivation were quite complementary. The research has been funded, since 1983, by the UK Joint Committee of the Economic and Social Science Research Council and the Science and Engineering Research Council, a committee set up to fund interdisciplinary work.

The research is aimed at analyzing case studies of failures and successes in structural engineering and using the results to develop a management tool which might provide some early warning of a likely failure. The method is based on the idea of developing a computer knowledge base which can be accessed through FRIL (Fuzzy Relational Inference Language) or through any other appropriate A.I. language. In simple terms, it is envisaged that the project engineer for some new structure will be able to use the computer to relate his proposals with the accumulation of past failures and successes held in the computer knowledge base. The measure of structural safety will be some measure of the similarity of his proposals with existing structures and its dissimilarity with failures.

## Research Methodology

Several case studies are currently in progress and one is completed (Pidgeon, Blockley and Turner 1986). The studies involve semi-structured interviews with key informants. The interviews are tape recorded and explore the background, work situation, communications network and work roles associated with a project. The first stage of the work is to establish a clear, qualitative account of the social and organizational context of each project. At the same time the chain of technical and social events leading up to a failure is established with particular attention being paid to the actions and decisions of the individuals involved, especially those which might be thought to involve the tacit or overt assessment of risk.

Since the enquiry is centrally concerned with failures, it is important to look for the unintended consequences of the actions and decisions of those involved. The information collected is then used to construct a "grounded theoretical" account of the socio-technical features of the case studies. Grounded theory (Glaser and Strauss 1968) is a technique which has been specifically devised to allow research workers to develop "middle range" social science theories from qualitative data. Turner (1981) has considerable experience with the method. For the successful "ongoing" case studies, information is collected at intervals to monitor responses of the personnel involved.

Emphasis has initially been on gaining access to <u>all</u> important personnel involved in a project. This includes client, designers (architects, structural engineers, services engineers etc.), contractors and sub-contractors and others (independent consultants advising on a failure, loss adjusters etc.). It is becoming increasingly difficult to secure such case studies. Any study involving current litigation is unavailable and any study older than about ten years involves "memory" problems. The "window" of suitable case studies is, therefore, rather small. The next stage of the work is therefore to study incomplete accounts (in the sense that not all important parties may have given their version of events).

It is worth emphasizing that engineers are not used to paying undue attention to research methodology because the study of physical systems has a fairly well established methodology. However, this sort of research owes more to social science than physical science and the methodology is more complex and thus requires much thought.

The Model

The knowledge base will be arranged hierarchically (Figure 1). At the lowest, most detailed level, are the stories of the case studies which are termed the "event sequence diagrams". Figure 2 illustrates just such a diagram taken from our first case study (Pidgeon, Blockley and Turner 1986). Figure 3 is a similar diagram for a colliery explosion study by Turner (1978). As concepts are clustered from the various case studies, new higher levels concepts derived; this is working from the bottom up. At the same time using the more general patterns described by Turner (1978), the model can be built from the top down. The area of difficulty is the middle ground where these two approaches must meet.

Top level rules might be as follows:

1. FRIL: Safe (x):- similar_exist_structs(x,y) and NOT similar_failures (x,z)

   English: x is safe if x is similar to existing structures y and it is not the case that it is similar to failures z.

2. FRIL: Similar_exist_structs(x,y):-safe(x,t) and schemes(x) and safe(y,t) and exist_structs(y) and similar(x,y)

   English: schemes x are similar to existing structures y at time t (to level $X_x$) if x is safe (to level $X_x$) and y is safe (to level $X_y$) and $X_x$, $X_y$ are similar.

3. FRIL: safe(x,t):- project_history_OK(x,t) and proj-inf(x,t) and [{current-paradigm(x,t) and NOT accept_margins_ eroded(x,t)} or {outside paradigm(x,t) and ideas_tested(x,t)}] and NOT pro-

ject_sensitive_to_use(x,t)

English: x is safe at time t if history of project is acceptable to time t and the project information is dependable and [{the ideas being used are in current paradigms and acceptable margins have not been eroded} or {the ideas are outside the current paradigms and the ideas have been well tested}] and it is not the case that the project is sensitive to use by persons after project completion.

4. FRIL: current_ paradigm(x,t):
-techniques_well_tried_tested(x,t) or [some technological-change(x,t) and change_tested(x,t)]

English: the ideas being used in x at time t are in current paradigm if they are well tried and tested or if there has been some change from well known techniques then the changes have been tested.

5. etc.

## Inference Mechanism

A new method of inference for knowledge based systems is being used (Blockley and Baldwin 1986). The method combines features of the theory of fuzzy sets, Shafer's theory of evidence and probability theory. The method follows the Horn clause logic representation of Prolog. Unlike probability theory it does not rely upon a closed world form of knowledge representation and it can allow the modelling of total uncertainty. The Prolog form of negation by failure is not used. The method involves the use of necessity and possibility measures for the truth or dependability of a proposition.

## Discussion and Conclusions

This paper describes an ongoing project which has a rather ambitious end. That is to produce a knowledge based computer system which might be an aid in the management of the safety of a project. However, the usefulness of social science based investigations of case studies of structural failures has already been demonstrated with immediate practical lessons being learned. Progress in interdisplinary research is bound to be slower than traditional research at first because of methodological problems. However, if the right conditions can be found it can be very rewarding. In my view, it is essential for, progress of research on the role of human error in structural safety.

References

Blockley, D.I. "The Nature of Structural Design and Safety", Chichester, Ellis Harwood, 1980.

Blockley, D.I. and Baldwin, J.F. "Uncertain Inference in Knowledge Based Systems" ASCE, (submitted for publication).

Turner, B.A., "Man Made Disasters", London, Wykeham, 1978.

Pidgeon, N.F., Blockley, D.I. and Turner, B.A. "Design Practice and Snowloading-Lessons From a Roof Collapse", The Structures Engineer, 64A, No.3, March 1986, 67-71.

Glaser, B. and Strauss, A., "The Discovery of Grounded Theory", London, 1968.

Turner, B.A. "Some Practical Aspects of Qualitative Data Analysis", Quality and Quantity 15, 1981, 225-7.

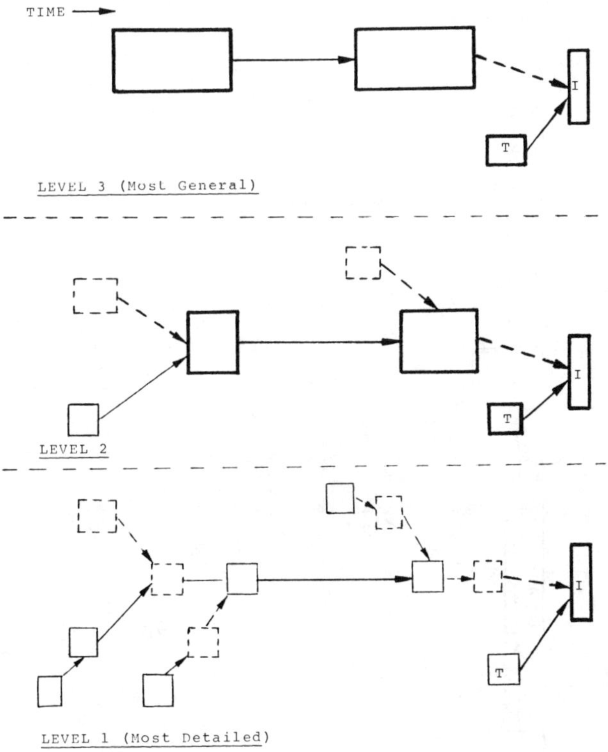


Fig. 1 Modelling Event Sequences.

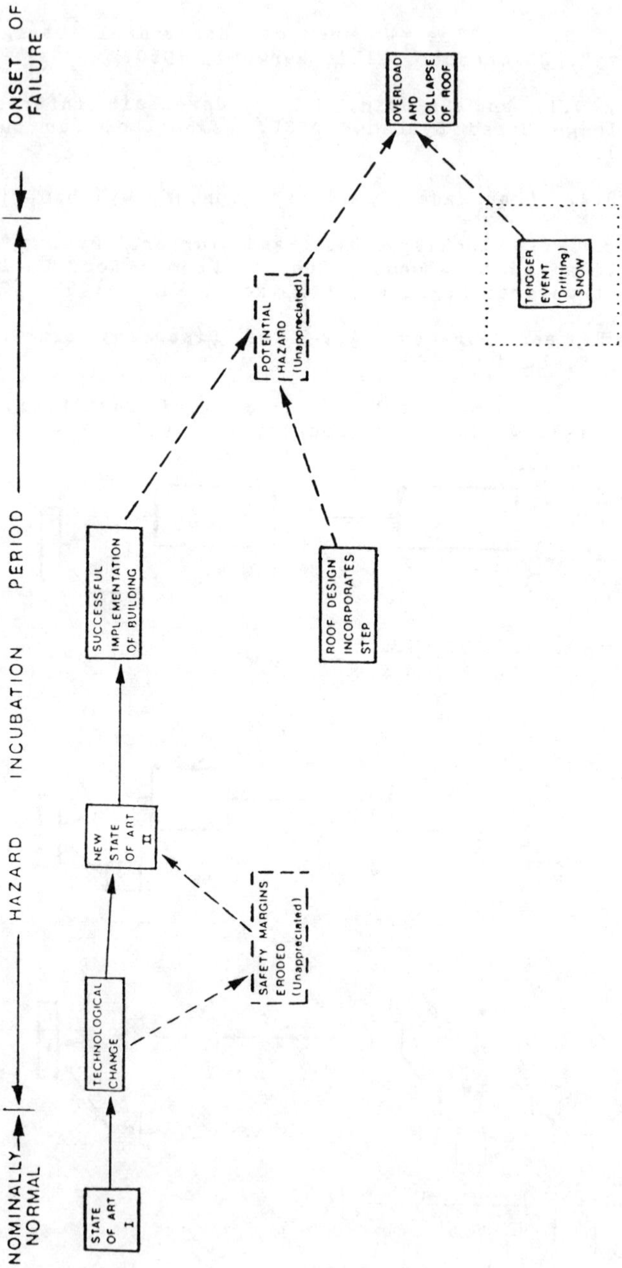
NOMINALLY NORMAL
HAZARD INCUBATION PERIOD
ONSET OF FAILURE
STATE OF ART I
TECHNOLOGICAL CHANGE
SAFETY MARGINS ERODED (Unappreciated)
NEW STATE OF ART II
SUCCESSFUL IMPLEMENTATION OF BUILDING
ROOF DESIGN INCORPORATES STEP
POTENTIAL HAZARD (Unappreciated)
TRIGGER EVENT (Drifting) SNOW
OVERLOAD AND COLLAPSE OF ROOF

Fig. 2

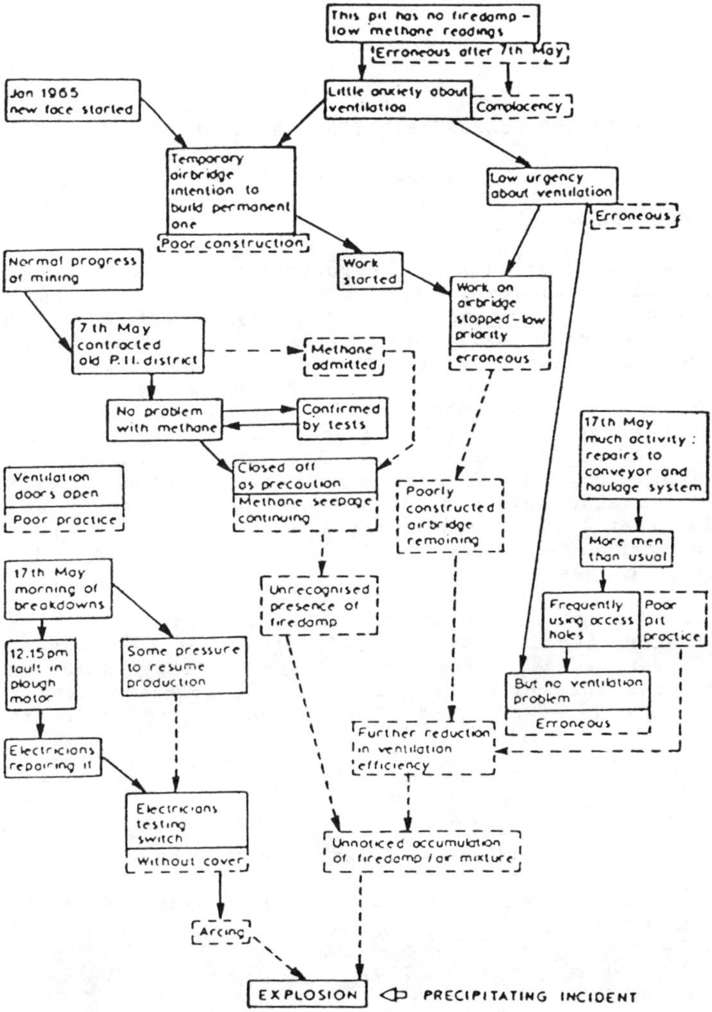


Fig. 3. Schematic Representation of the Situation at Cambrian Colliery immediately after 1240 hours on May 17, 1965.

# FORMATS AND RESPONSIBILITIES IN STRUCTURAL DESIGN

Janusz Murzewski*

## Introduction

Clear delimitation of responsibilities has reduced the number of accidents, e.g. in vehicular traffic and other fields where safety problems are very important and human errors are the most frequent causes of failures. There is, however, no crisp understanding of responsibility in structural safety (Blockley, 1983). Either gross errors or a major force are indicated in the conventional deterministic surveys about structural failures (Dmitriev 1953, Hammond 1956, Mitzel, Stachurski & Suwalski 1973, Augustyn & Sledziewski 1976). A major force, in the opinion of experts, sometimes also covers obvious faults or negligence as well as the fact that responsibilities in building have not been clearly specified and the inspection practices not adequately foreseen.

An implementation of probabilistic formats of design (Thoft-Christensen & Baker, 1982) etc., can make the situation even worse. The level-2 probability-based design introduces the reliability index $\beta$, defined as the reciprocal of the coefficient of variation of the random safety margin g ,

$$\beta = 1 / v_g = ( \sigma_g / \bar{g} )^{-1} \qquad (1)$$

where $\bar{g}$ = mean value, $\sigma_g$ = standard deviation.

If safety condition can be expressed by a simple inequality

$$A < B \quad \text{or} \quad g = B - A > 0, \qquad (2)$$

for A - random action (load), B - random resistance (capacity), the reliability index is expressed as follows

$$1 / \beta = \sqrt{\sigma_A^2 - 2 \rho \sigma_A \sigma_B + \sigma_B^2} / (\bar{B} - \bar{A}) \qquad (3)$$

where $\rho$ = coefficient of correlation,
$\rho$ = 0 if A and B are independent.

---

*Professor of Civil Engineering, Politechnika Krakowska, 31-155 Krakow, Poland

A probability of failure, q, for normal random variables A and B usually is calculated by means of the Laplace function,

$$q = \Phi\,(-\beta), \quad Q = 1 - q \tag{4}$$

A reliability condition

$$\beta > \beta^{*} \quad \text{or} \quad Q > Q^{*} \tag{5}$$

where $\beta^{*}$ and $Q^{*}$ = specified values (Fig.1), is sufficient for design of structures, but it gives no possibility, in case of an accident, to indicate who is responsible for the failure - the contractor (B is too small ?) or the user (A is too large ?). That is why, perhaps, so much attention is paid to design values X*, associated with $\beta$, (DIN) etc., although these are necessary in level-1 design formats but not on level-2.

Sometimes the safety condition cannot be defined in an explicit form (Dimitriev 1953). A function of many basic variables X1, X2, X3, ... Xn is formulated and it must be positive in the safe domain,

$$g = g\,(X1, X2, \ldots, Xn) > 0 . \tag{6}$$

The idea that a design value Xi* is specified for every basic variable Xi according to level-2 calibration procedures makes no progress in the problem of delimitation of reliabilities - in comparison with conventional methods. It hardly gives anything from the quantitative point of view, because standardization task committees often manipulate the data in such a way that they obtain the old values again.

A combination of the level-2 safety measure with design values is contradictory. The design values Xi* truncate their probability functions f (Xi) (Fig. 2). Nevertheless, the natural curves are used to evaluate q. Neglecting the effects of control is unjust and it does not encourage people to a greater care for the high quality of building and the avoidance of errors.

## Domains in the Space of Structural States

A pragmatic approach to the harmonization of structural design, construction and maintenance should be such that not necessarily the probability of failure, q, but a probability of rejection, w, should be taken into account,

$$w = 1 - \text{Prob}\,(X1 < X1^{*}, \ldots, Xi \lessgtr Xi^{*}, \ldots, Xn > Xn^{*}), \tag{7}$$

where the sign < is for such basic variables as an unfavorable load and the sign > is for mechanical properties.

$\Omega$ = 1 - w means probability that the structure has been used during its service life without objections. There are four ways to enhance the acceptance level $\Omega$ :

1) to make the structures stronger, i.e., to increase their means capacity $\bar{B}$;
2) to reduce the expected loads or other controllable actions $\bar{A}$;
3) to fix a lower level for the minimum capacity B*;
4) to augment the maximum tolerable loads A*.

The first two ways increase the costs and the other two ways reduce the safety margin, B* - A*, which is left for design errors. There is another way which seems to be more advantageous. It is a reduction of number n of the basic variables X1, X2, ...,Xn. The probability $\Omega$ increases because the probability density function, f, is integrated over a $(1/2^n)$-th part of the n-dimensional space of structural states,

$$\Omega = \int \cdots \int_{X1<X2^* \ldots Xn>Xn^*} f(X1,\ldots,Xn)\, dX1\ldots dXn \tag{8}$$

and the integral $\Omega$ will increase as n decreases.

Furthermore, the number $\binom{n}{2}$ of boundary regions is reduced if n is small and opportunities of human errors are the most dangerous just in those fuzzy regions.

The problem of responsibility would disappear if n=1. Therefore, the least number of state coordinates is 2, i.e., the demand A and the resistance B. In this case, A* = B* and no margin is left for design errors. That is why 3 dimensions of the state space have been suggested in Ref. (Murzewski, 1974). The third coordinate C in the space of structural states may be defined

as a correction factor,

$$C = B / A \tag{9}$$

or as an additive correction element,

$$C = B - A \tag{10}$$

C is treated as a random variable, as are A and B in a uniform probabilistic approach to structural safety. The random variable C depends on all those uncertainties which cannot be controlled by the constructor and the user. The designer is supposed to control the correction element C and it is a basis for a more detailed definition of his responsibility field. Neither the constructor nor the user should be authorized to change anything in the structural system with the agreement of the designer. Such legal situations already exist in many countries.

Enumeration of human errors given e.g. in Refs by (Ravindra, 1985, Matousek, 1982), shows that any one of the three responsible individuals, i.e., the designer, the constructor, and the user can commit them. Too little attention is given perhaps to errors of the user, both in theoretical papers and in building maintenance practice. The users have often no idea about safety factors and they do not care too much about the structural safety, e.g., they neglect removing a catastrophic snow from the roof of their buildings. In rare cases of irresistible elemental actions (earth-quake, flood, etc.) the user should be warned in advance and let the building be evacuated. So it is out of service and it must not be included in the number of unreliable objects which have failed during their service.

## Gross Errors and Accidental Situations

Minor human errors can cumulate and cause a structural failure. Theory of probability is useful in prediction of such random events. The question is whether gross errors can be treated as random variables as well. In the author's opinion they are rare random events but they should not be taken into account in frames of regular design situations, neither permanent nor transient. It would be better to treat them as accidental design situations. Probability q' of failure in effect of an accident may be much greater than it is in the regular design situations. It might be comparable with standard probability, q, of an ultimate limit state only after multiplication by the probability of occurrence, p, and a probability $1 - \psi$ that the gross error had not been discovered on time.

$$q' \; p \; (1 - \psi) = q. \qquad (11)$$

The structural analysis of an ultimate limit state is also different in the accidental design situations. Usually, in regular design situations, the elastic and stable behavior is required. Eventually some limited plastic and post-buckling deformation are admitted. Such tough requirements are rather unjustified from the economical point of view in the accidental situations such as fire, explosion, collision, etc. Cases of gross errors may be included to such rare accidents. Then a full plastic mechanism may be developed in the ultimate limit state and even a partial collapse of some less important parts of the structure may be admitted. However, the most important point is to assure the safe escape and survival of people, to prevent a progressive collapse and to mitigate the disaster. The designer should feel responsible for foreseeing the most likely accidents and gross errors. Elaboration of standard hazard scenarios, structural sensitivity analysis (Nowak, 1983), and error elimination models (Lind, 1983) may help for such predictions.

**Checking Value Optimization**

The probability $\Omega$ concerns a random event defined so that there are no objections relative to structural safety in the three domains of responsibility,

$$\Omega = \text{Prob} \quad (A < A^*, \quad B > B^*, \quad C < C^*). \qquad (12)$$

Definition (12) gives a clear delimitation of responsibility domains and it may be helpful in preventing human errors. On the other hand, it gives much freedom of control to the responsible individuals. They are free either to accept or to reject elements of more or less approximate structural designs ($C_i$), stronger or weaker structural materials ($B_i$), lighter or heavier loads applied on the structure ($A_i$) only if the more or less favorable elements compensate so that their limit values C*, B*, A* are not exceeded, respectively.

Admission of a fixed probability of objections, $w = 1 - \Omega$, may give a basis for calibration of the limit values A*, B*, C*. However, it would not be the best solution. A minimum cost condition from Ref. (Murzewski 1970) which takes all costs under consideration and the efficiency of control $\psi$,

$$K = K0 + (1 - \psi) K1 \, q + \psi K2 \, w = \text{Min}, \qquad (13)$$

shows that the hazard functions h(X) and more exactly - the hazard scales X h(X), so called, determine the optimal design values,

$$A^* h(A^*) = B^* h(B^*) = C^* h(C^*) = \text{Const}, \qquad (14)$$

where

$$h(A) = f(A)/F(A), \; h(B) = f(B)/[1 - F(B)], \; h(C) = f(C)/F(C)$$

$f(X) = dF/dX$ - probability density, $F(X)$ - cumulative function.

The hazard function h(X) is relative to the risk of statistical quality control of a random variable X. It is equal to the abscissa of the truncated probability distribution curve at the checking point X* (Fig. 2). There are also arguments from the theory of system reliability which indicate that the hazard function, h, and not the probability w should be the target measure in safety considerations.

The value 1/k of the constant on the right side of Eg.14 is strongly influenced by consequences of failure, i.e., by the safety class of the structure. Efficiency of control, size effects and time effects may easily be taken into account by an appropriate modification of the value 1/k. The relation of the responsibility delimitation problems with routine test procedures, inspection requirements and prevention of human errors need further research.

## Conclusions

Probability rules should not be misinterpreted in order to justify structural failures. The design format should be such that somebody always feels responsible for his safety domain and the domains should cover all cases of possible failures. The number of responsible individuals should be as small as possible. Three of them seem to be enough: the designer, the constructor, and the tenant. The design format should be harmonized with these conclusions. Separate probabilities wA, wB, wC, or better yet, the risks hA, hB, hC should be taken for specification of checking values of imposed actions, structural resistances and design errors. The efficiency of control may influence the specified values of safety measures.

## References


Augustyn J., Sledziewski E., "Failures of Steel Structures, (in Polish) , Warsaw, 1976.

Blockley D.I., "Reliability or Responsibility?," Structural Safety, Vol. 2, No. 4 (June 1983), pp. 167-175.

D I N, "General Principles on the Specification of Safety Requirements for Structures", Berlin-Koln.

Dmitriev F.D., "Structural Failures," Historical-Technical Surveys, (in Russian), Moscow, 1953.

Hammond R., "Engineering Structural Failures", London, 1956.

Lind N.C., "Models of Human Error in Structural Reliability," Structural Safety, Vol. 1, No. 3 (April 1983), pp. 167-175.

Matousek M., "Measures Against Errors in Construction," (in German), Basel 1982.

Mitzel A., Stachurski W., Suwalski J., "Failures of Concrete and Masonry Structures," (in Polish), Warsaw, 1973.

Murzewski J., "Safety of Building Structures," in (Polish) Warsaw 1970; (in German), Berlin, 1974.

Nowak A.S., "Sensitivity Analysis for Human Errors," ICASP 4 Firenze 1983, Proceedings Vol. 1, pp. 729-739.

Ravindra M.K., "Treatment of Gross Errors in Nuclear Plant Risk Studies - A Suggested Approach," ICOSSAR 85 -Kobe 1985, Proceedings Vol. II, pp. 41-50.

Thoft-Christensen P., Baker M.J., "Structural Reliability Theory and its Applications," Springer Verlag 1982."

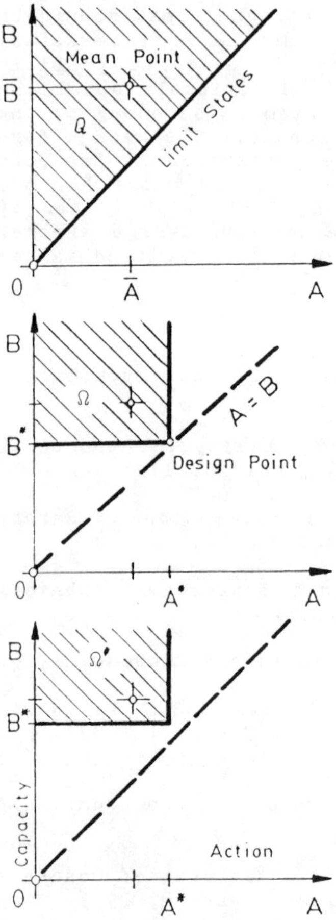


Fig. 1 Structural state domains

$Q$ = $P(A<B)$ no failure,
$\Omega$ = $P(A<A^*, B>B^*)$ no objections,
$\Omega$ = $\Omega_1$ $\Omega_2$ $\Omega_3$ no excessive actions nor unsatisfactory capacity nor non-tolerable design errors

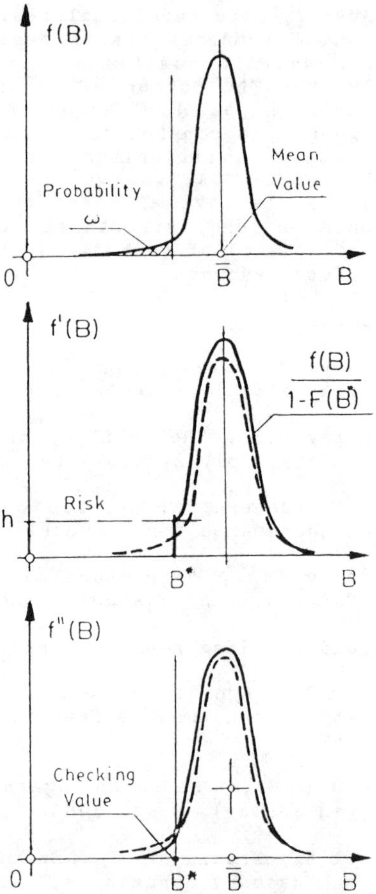


Fig. 2 Probability density curves of structural resistance

$f(B)$ - natural curve
$f'(B)$ - truncated curve,
$f''(B)$ - curve modified in effect of quality control

# GROSS ERRORS, SEISMIC MARGINS AND SEISMIC PRAS

M. K. Ravindra*, M. ASCE


## Abstract

The seismic risk and seismic margin assessments for nuclear power plants should consider the potential effects of design and construction errors. These errors generally modify the component responses and fragilities. The concern with design and construction errors is that they may simultaneously affect a large number of systems and structures and negate the possible benefits of redundancy. A methodology for treating the effects of gross errors in seismic PRA and seismic margin studies consists of enumeration and classification of errors, sensitivity studies to identify significant errors, collection of data on errors and incorporation into the margin and risk estimations.


## Introduction

Structures, equipment and systems in nuclear power plants are designed to withstand the effects of two levels of earthquakes (i.e., Operating Basis Earthquake and Safe Shutdown Earthquake) acting singly or in combinations with other postulated events. These design earthquake levels (OBE and SSE) are chosen such that they are larger than the levels historically occurring at the plant site. Design parameters such as the ground response spectrum, damping, material strength, and stress allowables are also selected conservatively. Therefore, nuclear power plants are expected to have seismic capacities much in excess of the SSE levels.

In recent years, a number of probabilistic risk assessment (PRA) studies of nuclear power plants have been conducted; many of these studies have included a consideration of the effects of seismic events. The key elements of a seismic PRA are seismic hazard analysis, seismic fragility evaluation of structures and equipment and analysis of plant systems and accident sequences. In the seismic hazard analysis, the frequencies of occurrence of different levels of earthquake ground motion are estima ted taking into account seismic history of the region around the site, capable faults and tectonic provinces in the region and their maximum earthquake potential, and attenuation of ground motion from the seismic source to the site.

---

*EQE Incorporated, Newport Beach, California 92680.

Seismic fragility is defined as the conditional probability of failure of a component (i.e., structure, equipment or system) for a given earthquake peak ground acceleration. Seismic fragilities are needed to estimate the conditional probabilities of accident initiating events (e.g., loss of off-site power, small loss of coolant accident, large loss of coolant accident, and reactor pressure vessel rupture) and the conditional probabilities of accident-mitigating systems (e.g., emergency core coo ling systems, reactor protection system, and quench spray system).

In the plant systems and accident sequence analysis, event trees are constructed to depict the progression of different accident sequences; fault trees are constructed to identify the different sequences of component failures that may lead to system failures. Fragilities are used to quantify these fault trees. The frequencies of occurrence of different earthquake levels are used to calculate the frequencies of occurrence of severe accidents (i.e., core damage, serious release, and unacceptably high consequences).

Some important conclusions of seismic PRAs are that seismic contribution to the plant risk could be significant, large uncertainties exist in the seismic risk estimates and that plant-specific analyses are needed to estimate seismic risk contributions and identify any critical weak links in the plant.

Although a seismic PRA would provide valuable insights in the seismic safety of a plant and identify dominant seismic risk contributors, the large uncertainties in seismic hazard at the site would make seismic safety decisions difficult if based only on seismic risk estimates. Therefore, interest in the nuclear industry has been recently focused on estimating the capacity of nuclear power plants to withstand earthquakes beyond the design basis levels. This is in response to an increased percep tion of seismic hazard especially in the Eastern United States. A methodology for performing a seismic margin study is being developed (Budnitz et al, 1985; Prassinos et al, 1985) for this purpose. A margin review earthquake level is specified and the objective is to assess if the plant seismic capacity exceeds this level. The plant seismic capacity, termed High Confidence Low Probability of Failure (HCLPF) capacity, is defined as the level of earthquake for which there is 95% confidence that the conditional probability of failure is less than 5%. Using the results and insights gained from past seismic PRAs, certain structures, equipment and systems in the plant are screened out as having generically high seismic capacities and/or not d ominant contributors to seismic risk. For the remaining components, the HCLPF capacities are estimated using a fragility evaluation. The HCLPF capacity of the plant is estimated by developing the Boolean expression for the relevant damage states an d quantifying them using component fragilities.

In the performance of the seismic PRAs and seismic margin studies, gross errors have not been considered explicitly; however, questions have been raised on potential gross design and construction errors (DCE) and their impact on seismic risk estimates and seismic margin insights. In these studies, fragilities of structures and equipment are generally estimated by extrapolating the SSE design-analysis to failure levels, taking into account the safety margins and their uncertainties. The fragility curves may be shifted to the left (i.e., increased conditional probability of failure at a given peak ground acceleration) if gross design and construction errors exist affecting the component (Figure 1). Therefore, the concern is that the results and insights gained from these studies assuming gross error-free plant systems and components may have been different if a potential for gross errors exists.

The role of gross errors in seismic PRAs and seismic margin studies is unique in that the plant is being reevaluated as opposed to being designed. Therefore, any benefits of quality assurance and quality control in reducing the likelihood of gross e rrors have already been accounted for in its design. The seismic capacity of the plant in its existing condition is evaluated in a seismic PRA or margin study. It is impractical, if not impossible, to reanalyze the plant structures and equipment for SSE response. However, the as-built conditions of structures and equipment may not be verified beyond a review of drawings and a cursory inspection for any obvious visible deviations. Testing for inplace material strength, and adequacy of anchorage is generally ruled out. Therefore, it is important to assess the rates of occurrence of gross errors and their effects on seismic risk and seismic margin estimates.

Types of Errors

A gross error is a deviation from standard practice which has an impact on plant safety or risk. For sorting out potential gross errors, we may start with all those deviations that may cause the value of the variable of interest (e.g., strength of a structural component) to lie outside its normal range of variation.

Gross errors could be random (i.e., occur at random in time and/or space) or systematic (persistent). An example of a random error is the use of regular bolts to anchor an equipment when the specification called for high strength bolts. An error in soil properties used in the soil-structure interaction analysis introduces systematic errors in the calculated responses at different component locations. Persistently poor quality of fabrication and errors in a computer code used are other examples of systematic errors. Errors could be binary (defective or not) or conti-

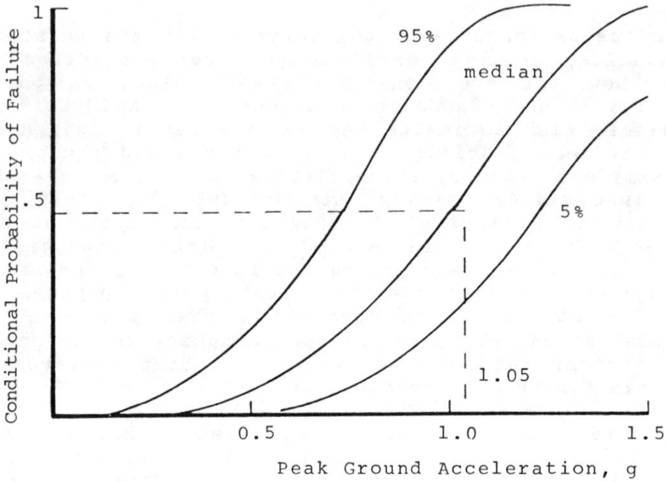


Fig. 1a Fragility Curves for a Gross-Error-Free Component.

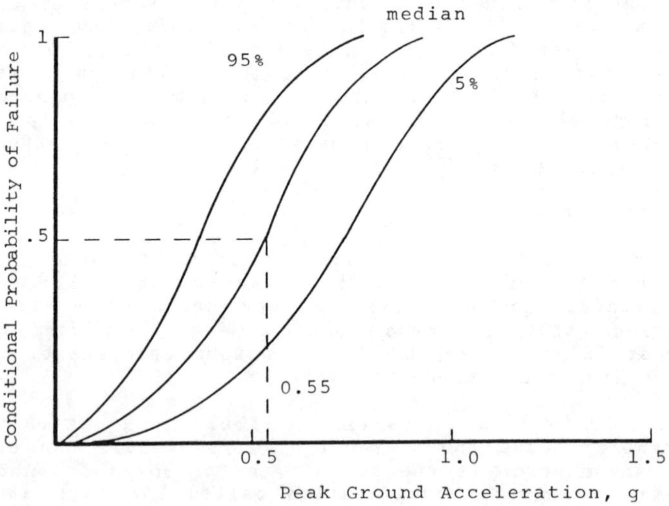


Fig. 1b Fargility Curves for a Flawed Component.

nuous (error in support spacing).

Another distinction between gross errors could be local and global: a local error affects only one component; a global error affects multiple components. Use of wrong size welds in one component is a local error; use of wrong soil properties in the seismic analysis is a global error.

Some examples of gross errors are given below.

Design Errors

- numerical (decimal point, conversion to different units, ommission of g)
- wrong assumptions (boundary conditions, loading conditions, wrong spectra, etc.)
- errors in material properties (steel, concrete, soil, etc.)
- errors in modeling (approximations, etc.)
- wrong mass (e.g., neglect of weight of water in tank)
- switching of drawings for Unit 1 and Unit 2
- mistakes in size (differences between model and drawings).

Construction Errors

- mistakes in size, material (bolt, steel)
- use of wrong type of fastener
- misreading of drawings, weld details, etc.
- defective installations: weld, anchor bolts, etc.
- reinforcement placement

State-of-the-Art Errors

- loads not envisioned during original design
- discovery of potentially active earthquake faults in the plant vicinity
- plants on soil sites not designed for soil-structure interaction effects
- safety-related components and systems designated as not safety-related
- use of materials susceptible to intergranular stress-corrosion cracking.

Sensitivity Studies

The risk-impact of a gross error may be assessed by postulating a specific error (i.e., type and magnitude) based on the significance of the failure mode and calculating the resulting change in the seismic risk (i.e., frequencies of core melt and ser ious release). Some preliminary studies on selected components at Zion (Ravindra, et al, 1984) have indicated that some components are immune to certain types of gross errors (e.g., column buckling strength is unaffected by improper installation of anchor bolts). In some instances, only errors of exceptionally large magnitude were found to significantly change the seismic risk predictions. By performing such studies on all the important structures and equipment in typical nuclear power plants, the types and magnitudes of significant errors may be identified. In a specific PRA, the approach could be to demonstrate either an absence of or an acceptably low rate of occurrence of such errors.

It should be noted that a lack of risk-sensitivity of some gross errors may have been caused by the shallow slopes and large uncertainties in seismic hazard curves. Hence, even moderate changes in seismic fragilities as a result of gross DCE may not significantly alter the seismic risk estimates. Sensitivity to gross errors may become more pronounced in seismic margin assessments.

A practical approach to the treatment of gross errors in the absence of data on their occurrence rates and magnitudes consists of sensitivity studies and use of expert judgment for categorizing errors and specifying reasonable ranges for their magnitudes. For each structure or equipment, the critical failure modes are reassessed as to their susceptibility to gross errors. Expert opinion is used in estimating the maximum magnitude of gross errors, and reduced seismic capacities of the component s (i.e., structure or equipment) are calculated. Using these reduced capacities, a revised estimate of the frequency of plant damage (e.g., core damage or serious release) or of the plant level HCLPF capacity is obtained. This revised estimate may still be acceptable. Otherwise, some limited sampling (i.e., review, reanalysis, inspection, and testing) may have to be performed to demonstrate that such errors do not exist or their rates of occurrence are not high enough to make these revised risk and margin estimates credible.

In performing such sensitivity studies, the importance of the particular component to the plant safety should be considered. If the component appears as a singleton "cut-set" for any significant damage state, a random gross error affecting this comp onent may have impact on the plant damage state frequency (or HCLPF capacity) so long as the other singleton "cut-set" components have much higher seismic capacities. Similarly, a "systematic" or global gross error that reduces the capacities of a number of singleton cut-sets is particularly important. On the other hand, redundant components (i.e., those that appear as "doubleton cut-sets" or "tripleton cut-sets", etc.) are affected more by systematic gross errors than random errors.

Incorporation of DCEs in PRA and Seismic Margin Studies

DCEs may modify the seismic responses and capacities of structural components and equipment. They may also introduce correlations between component fragilities, for example, through an error in the computer code or by a general poor quality construc tion. A possible way of including the potential effects of DCEs in a seismic PRA or margin study is to modify the component fragilities based on the rates of occurrence of DCEs and their magnitude distribution. Figure 1a shows the family of fragili ty curves for a component that is gross error-free. Figure 1b shows the fragility curves for the same component if there is a gross error. Such fragility curves may be developed by knowing the magnitude of the error and how it modifies the componen t response or capacity.

The fragility of a component may be written as:

$$f(a) \simeq \sum_i \sum_j r(i,m_j)\ f_i(m_j,a) + f_o(a) + f_R \qquad (1)$$

where:

$r(i,m_j)$ = rate of occurrence of error type i with magnitude $m_j$.

$f_i(m_j,a)$ = conditional probability of failure of the component with gross error type i and magnitude $m_j$ subjected to an earthquake of peak ground acceleration a.

$f_o(a)$ = conditional probability of failure of the error-free component.

$f_R$ = probability of failure of the component due to random failures (i.e., the unavailability due to maintenance or failure damage which is non-strength related).

It is envisioned that the occurrence rates and magnitudes of different significant errors may become available by following a well-planned program of data collection and that the fragilities of components are calculated using this information on DCEs . Note that the occurrence rates of errors of different types i and magnitudes $m_j$ are also uncertain. Hence probability distributions of error rates characterizing this uncertainty may have to be developed.

Some DCEs may affect more that one component simultaneously. For example, fragility evaluation in current seismic PRAs or margin studies is done by extrapolating the design-analysis information. If there was a design error in the structural analysis (e.g., selection of response spectra and parameters of the seismic model), the design responses of several components could have been in error. Hence, the fragilities developed on the basis of erroneous design responses are simultaneously in error for these components. A possible way to treat such a "systematic" or global error is to identify all the components that are affected by the error and calculate their revised capacities (Table 1). The fragilities of accident sequences involving these components are recalculated using the information on error occurrence rate and the revised seismic capacities of components.

Table 1, A Sample Table Showing the Influence of Systematic Errors on Component Capacities

| Error Type | Magnitude | Rate of Occurrence | Revised Median Ground Acceleration Capacity of Components as Fraction of Original Values | | | | | |
|---|---|---|---|---|---|---|---|---|
| | | | 1 | 2 | 3 | 4 | 5 | 6 |
| 1 | $m_1$ | $10^{-2}$ | 0.80 | -* | 0.70 | - | 0.80 | 0.90 |
| | $m_2$ | $10^{-3}$ | 0.60 | - | 0.70 | - | 0.75 | 0.85 |
| | $m_3$ | $10^{-4}$ | 0.60 | - | 0.65 | 0.95 | 0.70 | 0.80 |
| 2 | $m_1$ | $10^{-2}$ | - | 0.80 | 0.95 | 0.90 | 0.80 | 0.95 |
| | $m_2$ | $10^{-3}$ | 0.80 | 0.70 | 0.85 | 0.90 | 0.80 | 0.90 |
| 3 | $m_1$ | $10^{-2}$ | 0.75 | - | - | 0.80 | 0.95 | 0.95 |
| 4 | $m_1$ | $10^{-1}$ | - | 0.95 | 0.95 | 0.80 | 0.90 | - |

*Not affected by error

The above methodology is applicable for random errors of both local and global types. At present, only those gross errors that are considered misapplications of existing technology may be treated formally in seismic PRAs and seismic margin studies. The "state-of-the-art" errors are hard to identify. Once identified, they are corrected throughout the industry so that they may not recur.

Research Needs

There is apprehension in the nuclear industry that the results and insights gained from seismic PRAs and seismic margin studies may have been different if a potential for gross errors exists. No data exists that substantiate this apprehension. Also, no quantitative analyses have been performed to study the impact of gross errors because a rational methodology and the required data on occurrence rates and magnitudes of errors do not exist. Hence, a concerted research effort is needed

- to assess if DCEs are important in seismic PRAs and seismic margin studies, and if they are

- to develop a logical framework for data collection on DCEs and their rational treatment in seismic PRAs and seismic margin studies.

The first objective is achieved by reviewing nuclear operating data on failures through DCEs (e.g., license event reports, nuclear plant reliability data system, etc.) and the actual earthquake experience data to identify potential DCEs. Expert opinion is sought to postulate the magnitude ranges and rates of occurrence of DCEs. Sensitivity studies are performed to study the significance of DCEs in PRAs and margin studies.

The second objective of developing a methodology needs data collection on rates and magnitude distributions for different types of errors. Further work is needed in modeling the effects of state-of-the-art errors on seismic PRA results.

References

Budnitz, R. J., et al (1985), "An Approach to the Quantification of Seismic Margins in Nuclear Power Plants", NUREG/CR-4334, UCID-20444, August 1985.

Kennedy, R. P., and Ravindra, M. K., (1984), "Seismic Fragilities for Nuclear Power Plant Risk Studies", Nuclear Engineering and Design, Vol. 79, No. 1, pp 47-68, May 1984.

Prassinos, P. G., Ravindra, M. K., and Savy, J. B., (1985), "Recommendations to the Nuclear Regulatory Commission on Trial Guidelines for Seismic Margin Reviews of Nuclear Power Plants", NUREG/CR-4482, UCID-20579, December 1985.

Ravindra, M. K., Banon, H., Sues, R. H., and Thrasher, R. D., (1984), "Sensitivity Studies of Seismic Risk Models", Electric Power Research Institute, EPRI-NP-3562, Palo Alto, CA, June 1984.

Ravindra, M. K., (1985), "Treatment of Gross Errors in Nuclear Plant Risk Studies - A Suggested Approach", In Structural Safety and Reliability - Proceedings of ICOSSAR '85, Edited by I. Konishi, A.H-S. Ang and M. Shinozuka, Kobe, Japan May 1985.

# CONTROL OF HUMAN ERROR IN STRUCTURES

Niels C. Lind*

## Abstract

A main tenet of the structural engineering profession is that it can produce structures that will not fail. To maintain the truthfulness of this, most failures must be ascribed to human error. The profession must strive absolutely to reduce failure rates, but there is no public concern that failure rates are too high, nor public willingness to pay more, to get more structural safety. Failure rates must be lowered through more effecient checking processes: better error filters.

The scenario concept is explored briefly; it will require considerable development before it can become a useful tool of error control. Checklist, particularly in conjunction with computer aided design procedures, appear to have greater promise for immediate payoff, but only as a strategy specificly against errors of commission.

## Introduction

A mainstay of the structural engineering profession is the myth that we can build structures that will not fail. Just as the medical profession is supported by the myth that it can make the patients feel better - no one would go to the doctor if this were not so - so does the structural profession insist that it is in complete control, and that structures won't fail - in spite of the fact that they sometimes do. We maintain this stance of professional omnipotence at a cost: whenever a structure fails we are willing to sacrifice the individual firm, engineer or con-tractor to reaffirm that structural accidents don't happen but are created. The individual made a human error; the profession knows better. The notion that most structural failures are caused by human error is both dependent on and supportive of the idea that failures can be prevented. Indeed, can failures be prevented? In retrospect, every failure has been found to be preceded by a "chain of events" (actually a tree of faults) and could have been prevented, usually in many ways. Some failures are prevented everyday; most failures are continuously being prevented by monitoring performance. The process has the nature of a series of filters (Lind 1983).

---

*Department of Civil Engineering, University of Waterloo, Waterloo, Ontario, Canada, N2L 3G1

It is mildly paradoxical that each failure could have been prevented but all failures cannot. The heart of the paradox has to do with the uniqueness of the past versus the infinite multitude of features possible, with the resulting generic nature of prediction (fore-casting, scenarios), and that what actually happens has probability zero if described in sufficient detail. From this we learn the futility of "Monday quarterbacking". But we also learn that failure control must rest on an effective recall of past failure sequences embedded in the design process.

Risk is present in all activity. When the current process of research on human error in structures is completed (after perhaps 10 or 50 years), the best we can hope for is that there is a major reduction in the rate and severity of failures. This reduction will not be gratis -the necessary research and preliminary organization will be a sunken capital cost to society, and there will likely be an increased cost of error control for each structure. Error prevention, like crime prevention, is a matter of economic optimality.

No one has yet shown that the rates of structural failure are too high, notwithstanding that no failure is acceptable. Almost every failure teaches us that a small amount of extra care or effort could have prevented it - but the amount of effort put into design and inspection remains the same. The engineering effort put into a structure on the average is just the amount that we want to put into it- neither more nor less. It seems that, collectively, we are really quite satisfied with failure rates, as they are. We are quite proud of our low failure rates pointing out "structural failure" as an insignificant cause of death and higher insurance premiums. If we are in error here, it is because structures are too safe - we direct too many of society's resources to ensure structural safety; some of this could be reallocated into health screening, for example, to save lives and health care costs. This is quite true; however, the amounts involved are small, probably less than one percent of the cost of structural facilities, and the possible savings would not justify the increase in risk making, say, the house as risky as the car.

Every engineer can realistically hope not to be personally associated with a failure in their professional lifetime. We each put as much effort into our work as we judge is in balance, considering the benefits and risks involved. Failure rates are OK: small enough that the public won't question the myth of professional omnipotence and small enough that each of us can believe it won't happen to us. We cannot hope to have more resources allocated to better failure control. This means that we must improve reliab- ility by more efficient control of design, construction inspection and maintenance.

## Efficiency in Error Control

In the matter of reducing the rates of failure, what can be gained must be gained for free - or almost for free. We can eliminate any proposals that call for a general increase in the number of engineer-hours invested in each structure. The public will not be willing to pay for more safety where safety is perceived as complete, and engineers will certainly not want to absorb the cost of increased vigilance. The gains in structural safety must come about from better organization, better use of existing resources, or as a by-product of other improvements (e.g. computerization).

Errors are made frequently, but are most filtered out. The problem is to devise a better filter, or perhaps just an inexpensive, additional filter to stop those few errors that escape through the existing filters. A good approach to catch these escapees begins with understanding them. The last decade of research into the nature of human errors, in other fields (Swain and Guttman 1983) as well as in structural engineering (Melchers 1983) has taught us that human error is diversified. There are errors of negli-gence, of ignorance, of communication, etc. - indeed errors can be classified in many different schemes. Someone has even classified errors according to the list of cardinal sins. However different, these classifications all show the same feature: the errors that occur are strikingly different in endless variety. Indeed, those errors that stick in our minds have little in common with each other: Tay bridge, Titanic, Tacoma Narrows, ect. There is an endless list of failures from which we have learnt lessons, and each lesson is different. True, there is such a thing as the banality of structural failure (e.g. yet another case of "inadequate lateral bracing"), but it occurs on a level where, apparently, it cannot teach us very much. A major part of the problem, then, is that "human error" is an umbrella term, covering many ills that may each require a specific treatment if treatment is to be effective. There are no panaceas.

## Strategies of Error Control

Under these circumstances an inverse method of approach suggests itself: "Here's a procedure; what errors will it catch? And at what cost?" A catalog of such procedures can be ordered for a particular structural line of business according to efficiency. Efficiency is measured in terms of expected failure costs avoided, divided by the cost of the procedure. A firm or authority can then adopt as many of these procedures as appropriate and in order of decreasing efficiency, until the costs outweigh the perceived benefits. This list of candidates would include among others: Scenarios; independent approximate analysis; complete independent review of designs; expert systems; decision tables; and designer's checklists. Indeed, the list should include the entire repertoire of error control strategies, the subject of a separate section of this workshop.

The approach is pluralistic. To develop a fairly complete list of possible error control strategies, several approaches are possible. (1) A list of possible errors could be generated from the clauses in a design code - after all, each clause is there for a purpose which, supposedly, has something to do with promoting structural safety. The viewpoint here is that an error-free structure is one that conforms to the code in all respects. (2) Another approach is to ask structural design checkers (supervising engineers, building authorities etc.) what their strategies are; perhaps such strategies can be synthesized. (3) Alternatively, we can proceed from a list of known failures, and for each failure, determine the errors that "caused" it, and ask what would have been the best way to prevent it. This approach uses the principle of the _fault tree_: starting at the trunk of the tree (called the _top event_, representing the failure) unravel the root system, representing sequences of events that together cause the top event (Lind 1986).

Scenarios

The _event tree_ approach, in contrast to a fault tree starts with a plausible sequence of events (branches) unfolding from an existing state (the trunk). A _scenario_ explores just one path in an event tree. The scenario approach has been suggested as a subject to be explored in the search for better methods of quality assurance (Schneider 1985) and has been used effectively in the safety analysis of power plants. It is diffi-cult to ensure that a set of scenarios is complete, and their major advantage may be that they present a formal support for the use of imagination to find error sequences that haven't yet been observed in the real world. This is especially important for structures of a new type and of major importance, for which a failure would be a catastrophe. This includes nuclear power stations and large chemical plants. Dams, on the other hand, are probably not included because all failure modes and consequences are well known. Schneider (1985) tacitly acknowledges this practical difficulty with scenarios when he suggests the use of Turk-stra's rule to reduce loading scenarios to manageable loading cases in a systematic way for ordinary structures.

Checklists

The _checklist_ is an effective tool, quite specifically against errors of omission. It is employed in air transport and is indispensable for airline pilots and mechanics. Whether or not to use a checklist involves a balance between the complexity and permanence of the task as against the facility with which the checklist can be produced. Learning a simple standard procedure, such as to fly a single-engine aircraft or to design a structural member, involves in part the establishment of a mental checklist. As procedures become more complex, written checklists become more appropriate. On the other hand, as we move from structural members to a complete structure it becomes increasingly difficult to operate a fixed checklist; the things to remember vary according to circumstance. The checklist (in the sense of a list of things to do) for a complex unique structure is complex and is usually mentally established by the designer as the design proceeds.

Computer Controlled Design

The advent of the microcomputer has made it convenient for the designer to interact with the computer during the design process. The computer can prompt the designer as the design proceeds; sequences of such promptings (=checklists) can efficiently be assembled and tailored for the job at hand. Moreover, the designer's response can be recorded conveniently in the process. Errors of oversight or omission can be eliminated efficiently, i.e. without significant increase in the complexity of the procedure.

The conversational mode of computer-guided design can also be made to incorporate features of an expert system; guiding the selection of loading cases, for example. Schneider (1985) cites an intriguing case of a railway platform roof that was designed for snow, wind and snow plus wind - but failed because it was apparently not designed for wind acting on the snow piled on the roof. This failure is a curious vignette on established procedures which undoubtedly will become part of the lore of Swiss engineers for generations to come. However, for most of us who likely don't have to design such a roof in our lifetime, the lesson learnt is more general and less specific. It is possible for designers and checkers to overlook a weakness that snow and wind won't overlook. But the microcomputer (with multi-megabyte memory) is eminently suited to remember this kind of information when it is needed.

Appendix - Some Structural Accidents and Related Scenarios.

A <u>scenario</u> is a believable sequence of events. An accident is a sequence of events that actually happened and, therefore, is believable. One way to generate a scenario is to generalize accident sequences. For example, "on November 30, 1836 one of the spans of the chain pier at Brighton, England, was destroyed by a storm ... [drawings show] the sine curve undulations of the span in two segments with node at mid-span ... These drawings reveal a striking similarity to ... the undulations and collapse of the Tacoma Narrows Bridge" (Steinman and Watson 1941). Russell (1841) discussed "the general nature of the vibrations that destroy suspension bridges". Thus, in 1841 the principles of flutter instability of suspension bridges were already analyzed, tested and recorded (Steinman and Watson, 1941). In 1854 the great suspension bridge over the Ohio river at Wheeling was destroyed by wind in a similar mode.

Scenario A Structure is designed with inadequate stiffness, damping and aerodynamic stability and fails because of accumulated kinetic energy under a moderate steady wind.

Scenario B Profession fails to transmit the lessons learnt from the (repeated) failure of Brighton chain pier, spelled out by Russell (1841), leading to at least two repetitions (1854 and 1940).

Scenario C Designer overlooks a failure mode.

Each of the three scenarios above adequately describes the sequence of events that led to the failure of the Tocoma Narrows Bridge. Scenario A is very specific; as expressed in these terms the lesson has been learnt and this scenario is effectively prevented in future suspension bridges. Scenario B, on the other hand, places the onus on a general shortcoming of human organization ("History teaches us that we fail to learn from history"). The remedy is increased professional vigilance. Scenario C puts the onus on the designer, invokes professional responsibility and suggests checklists and other simple aids to organization as a remedy. This scenario, by the way, is quite general - The Tory Bridge and Titanic are both realizations of this scenario.

Reference

Lind, N.C., "Models of Human Error in Structural Reliability", Structural Safety, Vol. 1, No. 3, April 1983, pp. 167-175.

Lind, N.C., "Methods of Risk Analysis", Proc. of Conference on Hazardous Materials, IIASA, Laxemburg, Austria, July 1985.

Melchers, R.E. and Harrington, M.V., "Structural Reliability as Affected by Human Error", Proc. of Conference on Application of Statistics and Probability in Soil and Structural Engineering, Florence, Italy, June 1983.

Swain, A.D. and Guttman, H.E., "Handbook of Human Reliability Analysis with Emphasis of Nuclear Power Power Plant Applications", NUREG/CR-1278, U.S. Nuclear Regulatory Commission, Washington, D.C., Aug. 1983.

Steinman, D.B. and Watson, S.R., Bridges and Their Builders, McClelland & Stewart, 1941.

# THE CASE FOR CASE STUDIES IN ENGINEERING EDUCATION

Carl J. Turkstra*, F. ASCE

## Introduction

The sponsorship of this workship by the National Science Foundation suggests the existence of a major problem of "Quality Assurance" in the civil engineering profession. In many incidents of failure reported in press accounts and television reports, human error is a popular culprit. However, there are many complex social factors which lead to mistakes by individuals, poor quality and major organizational system failures.

Few would deny that the default by a northwestern utility on billions of dollars of publicly held bonds because of cost overruns and time delays represents a catastrophic engineering failure. Yet there were no identifiable human culprits. The space shuttle failure was technical in nature but organizational in its primary cause - the goals of the agency changed from technical ones to public relations and profit seeking. Engineers were told to stop being engineers and become managers.

The thesis of this paper is that a major cause of engineering failures is the inadequate education and limited talent of engineers. They have lost their position of leadership and hence control of their technical projects. Our educational system has encouraged this situation. As a corrective measure, the incorporation of formal case study techniques into civil engineering education is suggested.

## Recent Trends

Statistics now available indicate a decline in civil engineering enrollments in the United States. Among civil engineering faculty, there seems to be a widely held conviction that todays students are inferior to those of previous generations. Graduate schools and research are dominated by foreign students as Americans go off into other fields.

Closely linked to these trends are trends in practice. Whereas top levels of management in public agencies such as sanitation, water, highway and bridge departments, subway authorities etc. were once dominated by civil engineers.

---

* Professor of Civil Engineering, Polytechnic University, Brooklyn, New York

They are now controlled by lawyers, graduates in public administration and other generalists. In New York City which once had an outstanding technical civil service, there is not a single engineer among the commissioners surrounding the mayor. Recently a newly appointed commissioner of transportation announced major changes and "productivity" improvements overnight.

In New York we have a new subway tunnel under the East River - full of water. As a spokesman said - "all tunnels leak".

Our Wollman outdoor skating rink has become infamous. A trivial four month project involving a dent in the ground, grading, pipes and a concrete slab has taken five years to avoid completion at twice the original budget. The choice of a cooling system was made by a politician. Others have taken over a year to investigate.

In many respects, traditional construction companies have declined. One may now find a lawyer in a small office working with construction managers who control a myriad of subcontractors.

In consulting practice, designers have been isolated from the construction of their works. Design is often reduced to a secondary and rather mechanical part of the construction process. It took a midwestern judge to restate the obvious principle that licensing by the State as a professional engineer implies an irrevocable responsibility to ensure the quality of our work.

I would suggest that a major contributing factor to the decline of civil engineering is the modern trend in education. In the past three decades, our university programs have been modelled on requirements for the education of scientists and the human side of our profession has been downgraded.

The conflict between "scientific" reality and "political" reality has been discussed by many people. Through his novels, C.P. Snow described in some detail the clash between those who see the world in physical terms and insist on rational decision and those who see the world as a dynamic series of fleeting realities - the difference between those who seek optimal solutions and those who can work with much more complex variables. The latter group controls affairs and reap the rewards in every society.

By emphasizing science with little or no concern for liberal education, engineering schools deflect their graduates from real success and hence from control of their own works. In the present situation, engineering schools attract students with high math scores and, in many cases, limited verbal skills and limited intellectual interests.

Thus a viscious circle is set up where our graduates are seriously limited, they do not do well in life, and the quality of student attracted to engineering declines. As an equation one may write:

limited authority = limited prestige = poor remuneration = unattractive career = limited talent in the profession = poor quality = limited authority. (1)

Management Education - an Analogy

In its early stages several decades ago, management was considered an emerging science. The literature of the early 20s and 30s focussed on productivity and time-and-motion analysis. The search for "the one best way" to do human tasks was the primary objective.

To this day, success in such analysis and related management sciences such as operations research and systems analysis has not been rewarded with status, authority, or money in business or government.

The incredible success of business school graduates and their rise to dominance in human affairs is tied to their acceptance of soft sciences. It has been recognized that social factors, psychological factors and many other elements which can only be evaluated subjectively are as much reality as traditional hard science. An ability to deal with incomplete data, vague relationships and fuzzy objectives so as to exercise authority in an nonlinear transient world is accepted as a requirement for success in human affairs.

Many management schools attempt to prepare their graduates for this complex world through the use of case studies. In a typical prepared case, a management problem in a particular oranization is described briefly and related contributing elements in the situation are presented. Cases can be categorized by major classes of problems - e.g. marketing, productivity, labor relations, internal organizational conflict, Normally solutions are not suggested.

Students review the case presented, discuss the available information in the context of the material covered in specific courses they have taken and propose action. No unique solution is possible.

The benefits of such case studies are extensive. Complex factors with incomplete and subjective information are introduced and a variety of academic disciplines (e.g.psychology, organizational, theory, accounting, law) are linked. Perhaps most importantly, the need for judgment and character in selecting and defending an imperfect solution is emphasized.

The rewards for this style of education are also evident. The business schools attract intelligent and ambitious students. Graduates of good MBA programs are offered starting salaries often twice those of our masters' students. We work for them.

## The Approach in Legal Education

In the education and practice of law, the importance of cases is well known. Since human affairs are far too complex to be formulated in detail, legal disputes are generally resolved by extensions of precedent.

In legal practice, cases are categorized according to an hierarchy of basic concepts. Lawyers are trained to analyze a situation, identify the principles involved by a small set of key words and link these categorizations to a set of historical decisions. The art and science of a lawyer is the ability to develop supporting arguments from a subset of cases supporting his clients. Success in law requires an exceptionally fluent analytical mind coupled with imagination and verbal skills.

Recently, the president of a major university with a famous law school suggested that far too many intelligent young people go into law rather than engineering. Professor Salvadori of Columbia University was asked to comment on this observation based on his many years of experience as an expert witness and replied (at least partly in jest), "If the IQ of the average lawyer is 100, the IQ of the average engineer is about 60".

Fresh graduates from distinguished law schools have starting salaries as high as three times our graduates at any level. We work for them.

## Potential in Civil Engineering

It seems clear that the quality of engineering work is strongly dependent on the quality and authority of civil engineers. To attract quality students and recapture our traditional role in society, I believe we must rethink our educational system at a very fundamental level.

As Sir William Dawson said in 1854 during his inauguration as President of McGill University, "The success of a university is measured by the success of its graduates in later life". By this measure we are doing very poorly.

We must find a way to couple a scientifically oriented education with development of an ability to cope with poorly defined situations. Our graduates must hold their own with nontechnical peoples in terms of intellectual sophistication, verbal skills and social strategy. Otherwise, we will continue to loose control of our own works and become just another supporting group of technicians.

I believe one way to approach this problem is to introduce formal case studies in our curricula. These studies should couple technical questions with significant legal, economic, political, and social factors. The emphasis should be nonlinear in the sense that there is no path of deductive logic that leads to unique solutions.

Potential cases could involve, for example, budgeting operations in a highways department, planning operations in a mayor's office, site problems involving architects and public officials, contract disputes, labor disputes, or preliminary negotiations before a major project.

The methodology of case study could adapt the business school model. Each study should involve a few written pages outlining facts and suggesting major factors. Solution strategies (i.e. what should I do?) should be established by the students.

As in management, there must be an established methodology. This would include a set of key words with which to categorize a problem so students can answer the question, "what kind of problem is this?" There must also be a report format so that a student could concisely analyze the situation, propose a solution and defend his decision. The rules for preparing cases must be developed along with a new vocabulary.

As in management and law, a library of studies should be developed and set up formally as professional archives. There is no need to base cases entirely on facts and clearly the origin of any case and the individuals involved must be masked unless the details of a case are in the public record.

Based on the experience of other professions, the use of case studies can significantly contribute to the sophistication of our graduates and prepare them for more fruitful careers. As a long term investment, a systematic attempt to regain control of our works and reestablish the credibility and standing of our profession will go a long way towards preventing problems of quality assurance and human error.

# DESIGN CHECKING MODELS

R.E. Melchers* and M.G. Stewart**

## Abstract

Design checking has been suspected to be a major strategy in controlling human errors in design processes. The present paper reviews the context in which design checking models might be used and reviews some of the proposed models. It was found that the model which most closely fits independent checking survey results is an "S" curve similar to a "learning curve" used in educational psychology.

## Introduction

It is now quite clear that human error committed during the planning, design, construction and use of (civil) engineering projects is an immensely interesting topic, capable of immediately triggering off a rich variety of anecdotes amongst hardy practitioners and academics alike, yet one which is poorly understood in the objective sense. Little is known about how human errors come about or how they might be controlled or even prevented.

While it might be tempting to suggest that human error is inherent in all human activity and, therefore, cannot be controlled or eliminated, there is sufficient evidence that many (but not all) human errors are committed in the execution of relatively simple tasks (Matousek & Schneider 1976). Most errors are committed during the performance of such tasks. It is also clear that most errors are not due to unforeseeable events; most can be (or could have been) prevented, even without the benefit of hindsight. Since it is unlikely that anything can ever be done about truly unforeseeable events, a useful, if not very precise, restriction is to consider only errors involving (potentially) foreseeable events. A purist might like "foreseeable events" defined. However, a precise definition cannot be given at this time and may never be possible: it might be appropriate to record here that in the social sciences many concepts employed are not well defined and are also redefined as understanding of the concept grows.

---

*Professor, Department of Civil Engineering and Surveying The University of Newcastle, N.S.W., 2308, Australia
**Research Scholar, Department of Civil Engineering and Surveying, The University of Newcastle, N.S.W., Australia

The research work carried out over the last few years at Monash University (and now being continued at the University of Newcastle) has concentrated on human error in design processes for structural engineering (Melchers & Harrington 1982, 1984; Melchers & Stewart 1985). Unlike construction, design is a reasonably homogeneous process and hence more amenable, it was hoped, to systematic study. Design (and planning) has also been found to be a major activity in terms of human error occurrence (Matousek & Schneider 1976) and directly or indirectly has a bearing on errors which may occur in construction and in use of the structure (or other civil engineering projects).

The theoretical basis for the work has developed from a consideration of event-trees describing the decision and process steps in a typical project. Evidently, error initiation, error control and error consequence are major components. While these three components have been advocated as the major ones, [Kupfer & Rackwitz, 1980] the present work has shown that these components are not strictly separable, since, for example, some error control occurs during the very processes that govern error initiation. As a result, the authors' research has attempted to view the problem in considerable detail, with the aim of improving understanding the processes involved. Simplifications might be introduced at a later date if required for operational use of the human error models. An overview of the type of event-tree modelling that has been attempted has been given by Melchers & Stewart (1985).

Initially research by the authors' group concentrated on the description of error initiation or occurrence [Melchers & Harrington, 1982, 1984] as related to "micro-tasks" in structural design. Typical micro-tasks include algebraic calculations, table-look-ups, interpolation, etc. These tasks may be performed erroneously (error of commission) or not performed at all (error of omission). It is possible to have several other types of error, but these are generally of lesser importance [Melchers & Stewart, 1985]. Data on micro-task errors has been reported [Melchers & Harrington, 1982, 1984] however, these results must be seen as indicative since ambient conditions were not accounted for, nor was the data entirely based on responses from practising engineers. While these restrictions limit the validity of the conclusions, it is unrealistic at present to expect or even desire more precise information.

By setting-up an event-tree model of a specific macro-task, such as the determination of wind loading on a structure, or the design of one of its components, the micro-task data can be used to predict the error-rate distribution in performing the macro-task. This has been done and compared to available results for macro-task error rates [Melchers & Harrington, 1984]. Results of the event-tree modelling work have not yet been published. From preliminary assessment of the work it was felt that self-checking is an important component of the modelling. To date rather crude estimates have been used for the incorporation of self-checking effectiveness in the modelling procedure. However, it was felt that because of its importance, more attention should be given to self-checking effectiveness.

Before tackling self-checking it was considered appropriate to review models that have been proposed for independent checking (see below), as it was felt these might provide insight for self-checking, for which no models appear to be available at this time. The work described herein therefore relates mainly to independent checking of structural engineering designs.

## Design Checking

It is well-known that there are several potential strategies for control of design errors. Of these, the most commonly applied is design checking: it has been estimated that typically 0.5 - 1.0% of total project cost is spent on independent design checking by an in-house or external agent. Direct costs of local authorities, cannot, of course, be used because various forms of compliance verification are also carried out. Nevertheless, the question might well be asked how effective such checking is and what it achieves. Does it really ensure safer and/or better structures, or is it merely cosmetic, just as Bosshard (1979) has suggested that formal codified design has little to do with ensuring structural safety (or serviceability for that matter).

As noted above, understanding of the concepts of human error detection is essential if a complete model for human error in design is to be developed (Melchers & Stewart 1985). Such a model might be used in structural cost-benefit studies given various ambient conditions.

The work about to be described is based on small design tasks such as calculation, table look-up, etc. While these tasks are presently performed mostly by design engineers, it is increasingly becoming the case that computers will perform many of these tasks in the future. This does not significantly reduce the value of the investigations, however, since there will be many tasks still remaining, and these are likely to be similar or more complex in nature. Insight for simpler tasks will be needed before more difficult tasks can be modelled.

## Independent Checking

### Survey Instrument

Although some suggestions have been made regarding appropriate models for checking effectiveness, none of these is backed by empirical evidence. A first step was therefore to obtain some data.

So as to replicate real design office checking situations as clearly as possible, practising professional engineers were used as subjects through a mailed survey technique. While such a technique might alert prospective respondents about the nature of the exercise, and hence might well result in excessively good error detection rates, this was not considered to be of immediate importance. It was considered that the form of the model could be expected to be independent of the actual efficiency. The survey, which is described in more detail elsewhere (Stewart & Melchers 1985), was mailed to private consulting structural engineers, government agencies and semi-government agencies. One hundred and fifty sets of design sheets were sent out, with forty-seven useful responses being received.

Prospective respondents were asked to make note of any errors or mistakes in design loading calculations for a steel portal frame structure. Appropriate design criteria and information about the frame and its location etc. were provided. The respondents were asked to complete the task in about twenty minutes, but in any case to note the time that they spent on the task.

Within the format and the size of the survey task and its analysis, it was felt inappropriate to include significant errors of omission. Minor errors of omission would be detected, it was felt, in much the same way as errors of commission. The task consisted of ninety-three individual micro-tasks, of which nine were deliberately in error. Three of these were multiple step calculation errors, one was a table look-up error, two were errors of information transfer and a further three errors related to direction of the wind load pressures. These latter ones are of particular significance in wind load calculations, and relate directly to concept understanding.

Error detection by respondents was taken when a positive mark or correction had been noted on the design sheets as returned. If the respondent left no indication that a micro-task had been checked, it was assumed to signify acceptance of the result. Some respondents also detected "errors" where none were believed to exist. Such occurrences were regarded as irrelevant information for the purposes of this investigation.

Data Analysis

A histogram of the total time taken for task checking is shown in Figure 1. A Gamma distribution with a mean checking time of 20.6 minutes and standard deviation of 5.6 minutes reasonably fits the data. From the graph, it is evident that the lowest checking time was 10 minutes, but that the maximum time asymptotes to much higher values. Since there were ninety-three micro-tasks that required checking, the minimum time spent per micro-task was computed to be 6.5 seconds. For the average time, the time spent per micro-task increased to 13.3 seconds. The time spent per page of design varied from a minimum checking time of 3.3 minutes per page, to 6.9 minutes per page for average checking time.

While the above figures offer indicative times, such values would be expected to vary considerably depending on the complexity of the design to be checked.

Figure 2 shows the data points for checking efficiency (p) plotted against time (t) as obtained from the survey. A checking efficiency of unity was taken to imply that all errors were detected, while a value of 0 indicates that no errors were detected. The data points show a lot of scatter, as might be expected with the lack of control over test conditions. They do not indicate any obvious relationship, but do indicate a trend, namely that error detection increases as checking time increases.

It has been suggested (Kupfer & Rackwitz 1980) that error detection is related to search theory and can be expressed as a function of time by a negative exponential curve:

$$\bar{p}(t) = 1 - \exp(-\alpha t) \qquad (1)$$

where $\bar{p}(t)$ was the average checking efficiency for a total checking time t. The constant $\alpha$ may be assumed to be proportional to the level of detail examination and to the characteristics of the checker, and inversely proportional to the task size. As shown in Figure 2, this model, when compared with the data points, fails to fit the data adequately. Adjusting the value of $\alpha$ did not significantly improve the fit. Empirical evidence to support this model has been provided by Harris & Chaney (1969), who reported on inspection accuracy of electronic module assemblies. It is clear from Figure 2 that the trend of the experimental points does not pass through the origin, as the negative exponential rule (1) would suggest. Evidently, design checking is more complex than the simple inspection of components or the tasks to which search theory applies. However, it would be expected that once familiarization with a design had occurred, the process of design checking would have some similarity to search tasks and this is also evident from Figure 2.

A better description of design checking as a function of time spent is possible by applying the concept of learning. In the process of design checking, the designer first of all has to learn what is occurring within the design. Only after this has taken place can the checker endeavor to respond to any incorrect micro-tasks. Search theory then becomes relevant. In the field of the psychology of education, Hull (1952) postulated a learning theory. He stated that the probability of responding to a correct stimulus rather than an incorrect one increases slowly at first, then more rapidly, and then slowly again. A curve with this general shape was called a "learning curve", or a "sigmoid" because of its S-shape. Verification of the S-shaped learning curve was further provided by Estes (1959) who gave an appropriate formulation as well.

In terms of design checking, the use of an S-shaped curve has some appeal. The initial growth may be attributed to the designer attempting to understand the design concept and procedure. This is followed by a period involving checking of each micro-task for any errors, and in which many of the errors would be detected. Finally, the designer would reach the stage of diminishing returns, resulting in a reduced rate of checking efficiency. An S-curve which seems to be of the correct general form is:

$$\bar{p}(t) = 1/[1 + A \exp(-Bt^{1/2})] \qquad (2)$$

where A must be sufficiently large to ensure $\bar{p}(t) = 0$ when $t = 0$, and B is inversely proportional to the task complexity and proportional to the expertise of the design checker. The model expressed by eqn. (2) is compared with the survey data points in Figure 2 and is seen to provide a reasonable fit. However, it is important to note that many other possible S-curves could be postulated.

It is also possible to use a negative exponential curve shifted by a time amount $t_o$ (see Figure 2):

$$\bar{P}(t) = 1 - \exp[-\alpha(t - t_o)] \qquad t > t_o \qquad (3)$$

and this is also seen to be a reasonable fit to the data.

Evidence supplied by Harlow (1959) indicates that in education, learning curves progressively change from S-shaped to negative exponential as the subject's training increases. This observation is also of relevance here since a superior checking efficiency would be expected from an engineer with relevant expertise or experience in similar designs. It might be expected, therefore, that for the shifted negative exponential model (eqn. (3)), the term $t_o$ would reduce with experience. However, a truly negative exponential curve with $t_o = 0$ would be suitable only if the checker was extremely experienced in the particular aspect of his profession for which he carries out the design checking. This is seldom the case. It is more likely that most checkers have a broader-based knowledge. In addition, of course, each design is unique, so even an expert checker would require some effort to familiarize himself with the design. A learning curve, therefore, appears more appropriate than the negative exponential curve thus far suggested in the literature.

The checking efficiency as a function of error magnitude for each micro-task was also considered. This is plotted in Figure 3. This indicates that larger errors were more easily detected than smaller ones, as might have been expected. Lind (1983) has incorporated this expectation into the negative exponential model (1) to give checking efficiency as:

$$\bar{p}(x,t) = 1 - \exp\,[-\ h(x)\ \ c_o(t - t_o)] \qquad (4)$$

modified to allow for $t_o$ as in eqn. (3). The function h(x) was assumed to be a quadratic function of error magnitude (x) and $c_o$ is a constant. For a given time of checking, t, and with $h(x) = x^2$, expression (4) has been fitted to the data points of Figure 3 as shown. The curve asymptotes unite rather quickly but do not fit the data well. An improved fit is obtained if $h(x) = x^{1/5}$ is assumed.

Equation (2) may also be modified for error magnitude to give

$$\bar{p}(x,t) = 1/[1 + A\ \exp(-\ h(x)\ B\ t^{1/2}) \qquad (5)$$

where for given data values an error magnitude function

$$h(x) = x^{1/10}$$

appears to give a reasonable fit to the data. In both cases more empirical and theoretical justification appears desirable.

If instead of percentage micro-task error, the effect the error has on the outcome of the calculations is used (in this case a calculated bending moment), the results shown in Figure (4) are obtained. Equations (4) and (5) with the modified h(x) functions again produce a better fit than does the original quadratic function.

## Self-Checking

Self-checking by designers is a rather more difficult topic with which to deal. It is not possible to use a direct survey technique; review of the computations submitted by designers for previous surveys yielded insufficient information to be useful. Some very preliminary data has been obtained, however, from a review of first year engineering student examination papers in basic structural and dynamics topics. Some three hundred papers for each of the two separate subjects were scanned. Any changes made to written answers were recorded and an estimate made of the size of the original error. The results obtained were corrected using previous data on rate of error occurrence for a given size of error (Melchers & Harrington 1982). The corrected results are shown in Figure 5 as a histogram of percentage error magnitude.

The precise reasons for this observation are still under investigation; however it may well be that the larger errors are not detected at the time of computation or very shortly afterwards because the person concerned is unable to sufficiently divorce himself from the work in hand to take a more general overview. This phenomenon is well known, and raises interesting questions about what we really mean when we refer to "self-checking".

## Discussion

The work outlined above must still be seen as rather tentative first steps in our attempts to understand the nature and the control of human errors in design processes. As noted earlier, it is hoped that by understanding various elemental human error aspects, it will be possible to build-up a model which will help in estimating the importance of human error for a realistic design process.

The methods employed to date can be criticized on a number of grounds (Stewart & Melchers 1985); the important point, however, is that we cannot hope to proceed in our understanding unless we make an effort, even under quite imperfect conditions.

## Conclusion

The results given herein indicate trends that, to some extent, appear quite obvious. However, no data has previously been collected on error detection in design tasks and used to develop theoretical models. The models described herein are somewhat different to those proposed earlier without the benefit of empirical data.

## Acknowledgements

The work reported herein was supported by the Australian Research Grants Scheme, under grant no: F81 15110.

## References

Bosshard, W., (1979), Structural Safety - A Matter of Decision and Control, IABSE Surveys, S-9/1979, pp. 1-27.

Estes, W.K., (1959), The Statistical Approach to Learning Theory, (in) Psychology: A Study of Science, Vol. 2, General Systematic Formulations, Learning and Special Processes, Koch, S. (Ed.), New York, MCGraw-Hill, pp. 380-491.

Harlow, H.F., (1959), Learning Set and Error Factor Theory, (in) Pcychology: A Study of Science, Koch, S. (Ed.), New York, McGraw-Hill, pp. 492-537.

Harris, D.H. and Chaney, F.B., (1969), Human Factors in Quality Assurance, New York, Wiley.

Hull, C.L., (1952), A Behavior System, New Haven, University Press. Kupfer, H. and Rackwitz, R., (1980), Models for Human Error and Control in Structural Reliability, IABSE Anniversary Congress, Vienna, Austria, pp. 1019-1024.

Lind, N.C., (1983), Models of Human Error in Structural Reliability, Structural Safety, Vol. 1, pp. 167-175.

Matousek, M. and Schneider, J., (1976), Untersuchungen zur Struktur des Sicherheitsproblems bei Bauwerken, Institut fur Baustatik und Konstruktion, ETH Zurich, Bericht No. 59, Birkhauser Verlag, Basel and Stuttgart.

Melchers, R.E. and Harrington, M.V., (1982), Human Error in Simple Design Tasks, Department of Civil Engineering, Monash University, Research Report No. 3/1982.

Melchers, R.E. and Harrington, M.V., (1984), Human Error in Structural Reliability - I. Investigations of Typical Design Tasks, Department of Civil Engineering, Monash University, Research Report No. 2/1984.

Melchers, R.E. and Stewart, M.G., (1985), Data-Based Models for Human Error in Design, Fourth International Conference on Structural Safety and Reliability, Kobe, Japan, Vol. II, pp. 51-60.

Stewart, M.G. and Melchers, R.E., (1985), Human Error in Structural Reliability - IV: Efficiency in Design Checking Research Report No. 3/1985, Department of Civil Engineering, Monash University.

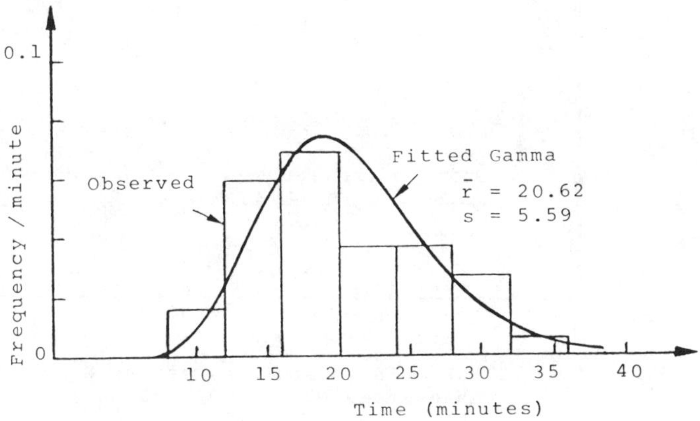


Fig. 1 Histogram of Total Checking Time Taken.

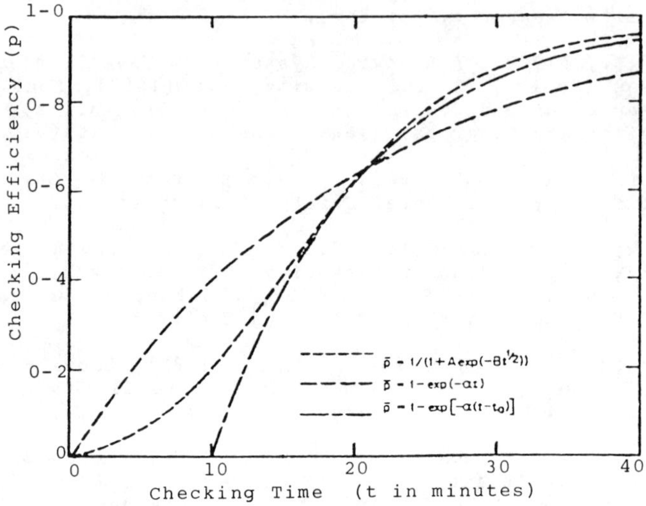


Fig. 2. Comparison of Error Detection Models

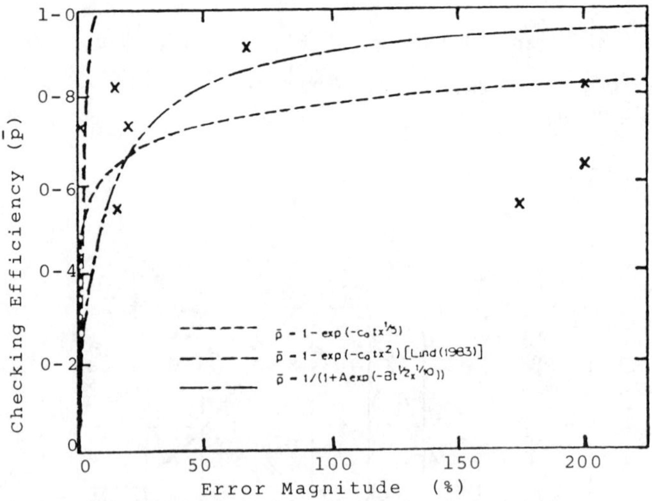


Fig. 3. Checking Efficiency versus Error Magnitude at t = 20 minutes

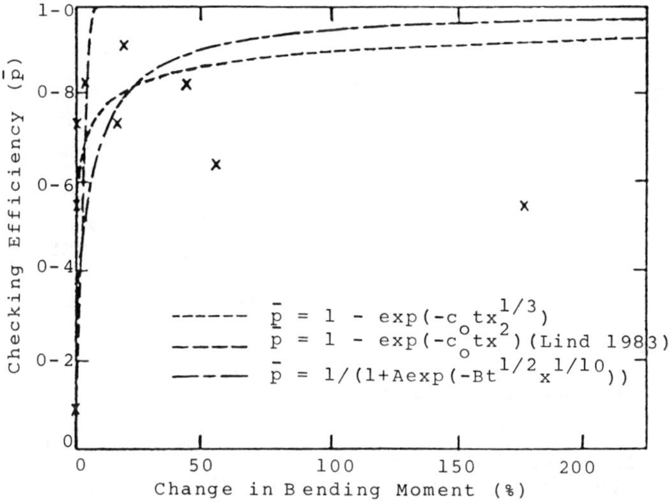


Fig. 4. Checking Efficiency versus Change in Bending Moment at t = 20 minutes

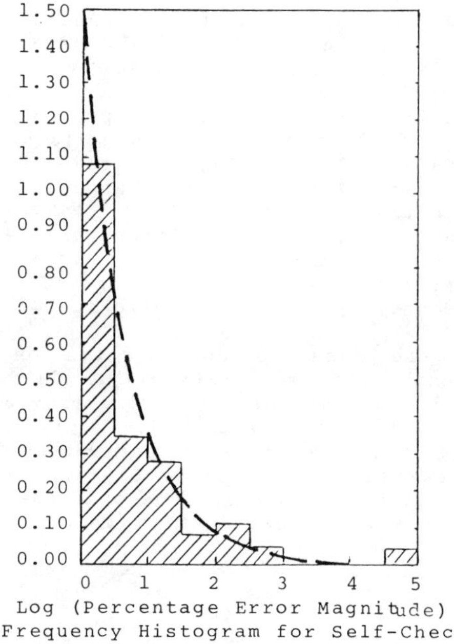


Fig. 5. Frequency Histogram for Self-Checking

# COMBINING H.E. IN ENGINEERING RISK ASSESSMENT

Dan M. Frangopol*, M. ASCE

## Abstract

In this paper an attempt is made to combine human errors in structural engineering risk assessment. The paper also includes results with regard to the sensitivity of the reliability index to changes in (a) the type of individual errors, (b) the type of combination of errors, and (c) the probability of occurrence of such errors.

## Introduction

It has been recognized that most of the structural failures are due to human errors (Allen 1981, Allen 1984, Blockley 1980, Frangopol, Nakib and Trautner 1985, Frangopol and Trautner 1985, Knoll 1982, Knoll 1984, Matousek 1981, Melchers and Harrington 1983, Nowak 1979, Nowak and Carr 1985). The goal of research on human errors has been defined by Allen (1984): "it is to help prevent failures due to human error". In this direction, significant contributions have been made by Allen (1981), Blockley (1980, 1981), Knoll (1982, 1984), Lind (1983a, 1983b), Matousek (1981), Melchers and Harrington (1983), Nowak and Carr (1979, 1985), Rackwitz (1977), Rosenblueth (1984), and Schneider (1981), among others. The recent advancements in the area of human errors have been summarized in the Proceedings of the Symposium on Structural Technology and Risk (Grigoriu 1984).

Some of the research concentrates on modelling the effects of human error mathematically. This is an extraordinarily complicated process involving many parameters (e.g. Knoll (1982) listed 71 parameters). Various simplified models involving a few very basic parameters have been suggested by Rackwitz (1977), Nowak (1979), and Lind (1983), among others. These models include only one error, with the exception of the error combination model proposed by Lind (1983a).

---

*Associate Professor of Civil Engineering, University of Colorado, Boulder, Colorado 80309-0428.

Using a simple discrete error model associated with a reliability index (beta) format, this paper describes an analytical investigation carried out to study the influence of combination of errors on the level of risk of failure. In order to accomplish this, both conservative and unconservative errors in the uncertainties underlying the resistance and load effect are combined. The results could be used to evaluate the significance of various combinations of errors on structural reliability.

## Discrete Model with Individual Errors

**Error-Free Reliability Index.** - Consider the fundamental case of structural reliability (i.e., one member - one load - one limit state) in which we assume that the resistance $R_o$ and the load effect $S_o$ are error-free known random variables. The error-free reliability index is given by

$$\beta_o = (\bar{R}_o - \bar{S}_o)/[\sigma^2 (R_o) + \sigma^2 (S_o)]^{0.5} \quad (1)$$

or

$$\beta_o = (\theta_o - 1)/[\theta_o^2 \quad V^2(R_o) + V^2(S_o)]^{0.5} \quad (2)$$

in which $\bar{R}_o$ = the mean value of $R_o$, $\bar{S}_o$ = the mean value of $S_o$; $\theta_o$ = the central safety factor (i.e., $\bar{R}_o/\bar{S}_o$), $\sigma(R_o)$ = the standard deviation of $R_o$; $\sigma(S_o)$ = the standard deviation of $S_o$; $V(R_o)$ = the coefficient of variation of $R_o$ (i.e., $\sigma(R_o)/\bar{R}_o$; and $V(S_o)$ = the coefficient of variation of $S_o$ (i.e., $\sigma(S_o)/\bar{S}_o$ . If $R_o$ and $S_o$ are independent normal variates, then the error-free probability of failure is

$$P_o = 1 - \Phi(\beta_o) = \Phi(-\beta_o) \quad (3)$$

in which $\Phi(.)$ = the standard normal probability distribution

**Combination of Errors.** - As Rackwitz (1977), Nowak (1979), and Lind (1983b) have pointed out, the occurrence of errors may change the error-free density functions of the resistance $f_{Ro}$ and of the load effect $f_{So}$. For example, if a different strength of concrete is used for a column, then the load carrying capacity R may have the density functions $f_{R-}$ (i.e., reduced capacity) or $f_{R+}$ (i.e., increased capacity) instead of $f_{Ro}$, as shown in Fig. 1(a). As another example, suppose that an error is made in the use of a structure. Then, the load effect S may have the density functions $f_{S+}$ (i.e., reduced load effect) or $f_{S-}$ (i.e., increased load effect), as shown in Fig. 1(b). The symbols o, -, and + in Fig. 1 indicate error-free, unconservative errors (i.e., decreasing the error-free reliability index), and conservative errors (i.e., increasing the error-free reliability index), respectively. As shown in Table 1, there are eight possible cases of combination of errors in R and/or S, that will change the value of the error-free reliability index $\beta_o$. The combinations denoted by B,C,F,

TABLE 1. - Combination of Errors

| Strength | Load Effect | Error Combination |
|---|---|---|
| $R_+$ | $S_+$ | A |
| $R_+$ | $S_o$ | B |
| $R_o$ | $S_+$ | C |
| $R_+$ | $S_-$ | D |
| $R_-$ | $S_+$ | E |
| $R_-$ | $S_o$ | F |
| $R_o$ | $S_-$ | G |
| $R_-$ | $S_-$ | H |

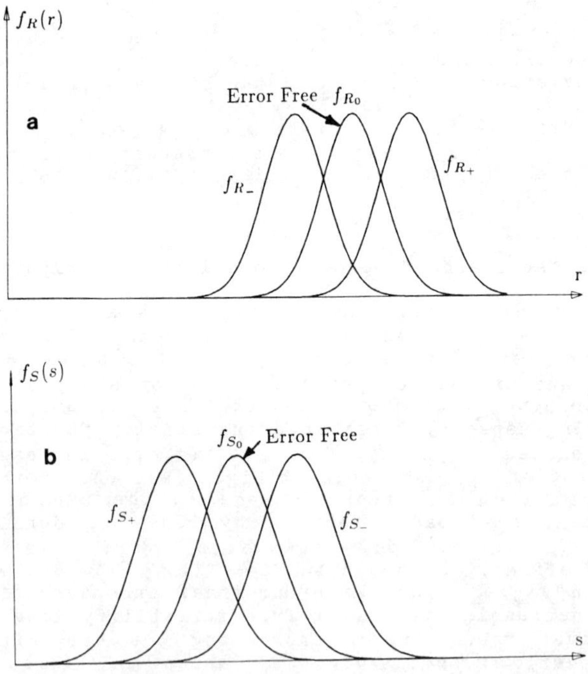


Fig. 1 - Additive Errors: (a) in Resistance; (b) in Load Effect

and G include only one error (i.e., errors in R or S), and the combinations denoted by A,D,E, and H include two independent errors (i.e., errors in R and S).

The parameters of the uncertainties underlying the two basic random variables R and S may change due to human error, as shown in Figs. 1(a)-(b). Two types of errors are considered in this paper: (a) additive errors (i.e., change only the mean value of a random variable, as shown in Fig. 2), and (b) multiplicative errors (i.e., change both the mean value and the standard deviation of a random variable, as shown in Fig. 3).

**Individual Additive Errors.** - Assume that an additive error occurs and modifies the error-free mean resistance $\bar{R}_o$ by a positive factor $F_R'$ (see Fig. 2). Then the error-free reliability index $\beta_o$ changes to:

$$\beta_R' = (F_R' \cdot \bar{R}_o - \bar{S}_o)/[\sigma^2(R_o) + \sigma^2(S_o)]^{0.5} \tag{4}$$

or

$$\beta_R' = (F_R' \cdot \theta_o - 1)/[\theta_o^2 \cdot V^2(R_o) + V^2(S_o)]^{0.5} \tag{5}$$

If $F_R' < 1$ the error is unconservative (i.e., reduces the mean capacity), and if $F_R' > 1$ the error is conservative (i.e., increases the mean capacity). Also, assume that the additive error in strength occurs with probability $P_{ER}'$. Then the total probability of failure including the additive error in strength can be shown to be

$$P_R^* = (1 - P_{ER}') \cdot P_o + P_{ER}' \cdot \Phi(-\beta_R') \tag{6}$$

or alternatively, the total reliability index is given by

$$\beta_R^* = \Phi^{-1}(1 - P_R^*) \tag{7}$$

where $\Phi^{-1}(.)$ is the inverse of the standard normal probability distribution.

If the additive error modifies the error-free mean load effect $\bar{S}_o$ by a positive factor $F_S'$, then the error-free reliability index $\beta_o$ changes to:

$$\beta_S' = (\bar{R}_o - F_S' \cdot \bar{S}_o)/[\sigma^2(R_o) + \sigma^2(S_o)]^{0.5} \tag{8}$$

or

$$\beta_S' = (\theta_o - F_S')/[\theta_o^2 \cdot V^2(R_o) + V^2(S_o)]^{0.5} \tag{9}$$

If $F_S' < 1$ the error is conservative (i.e., reduces the mean load effect), and if $F_S' > 1$ the error is unconservative (i.e., increases the mean load effect). Also, assume that the additive error in load effect occurs with probability $P_{ES}'$. Then the total probability of failure including the additive error in load effect can be shown to be

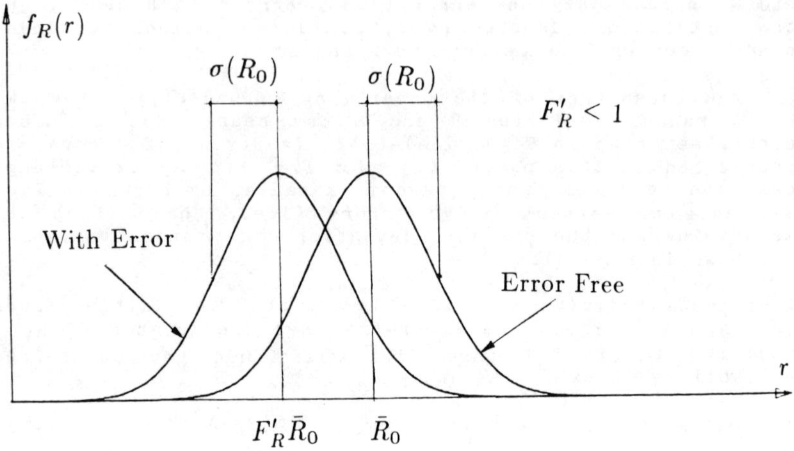


Fig. 2 - Additive Error in Resistance

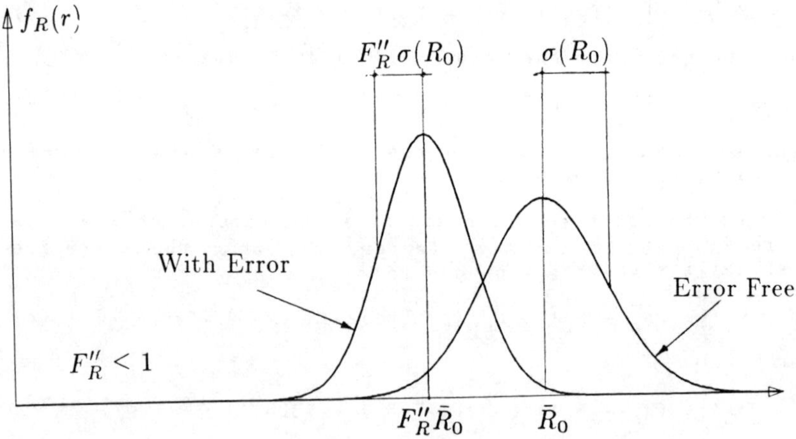


Fig. 3 - Multiplicative Error in Resistance

$$P_S^{*} = (1 - P_{ES}') \cdot P_o + P_{ES}' \cdot \Phi(-\beta_S') \quad (10)$$

or, alternatively, the total reliability index is given by

$$\beta_S^{*} = \Phi^{-1}(1 - P_S^{*}) \quad (11)$$

Individual Multiplicative Errors - If a multiplicative error occurs and modifies both the error-free mean resistance $R_o$ and the error-free standard deviation $\sigma(R_o)$ by the same positive factor $F_R''$ (see Fig. 3), then the error-free reliability index $\beta_o$ changes to

$$\beta_R'' = (F_R'' \cdot \bar{R}_o - \bar{S}_o)/\{[F_R'' \cdot \sigma(R_o)]^2 + \sigma^2(S_o)\}^{0.5} \quad (12)$$

or

$$\beta_R'' = (F_R'' \cdot \theta_o - 1)/\{[F_R'' \cdot \theta_o \cdot V(R_o)]^2 + V^2(S_o)\}^{0.5} \quad (13)$$

Also, assume that the multiplicative error in strength occurs with probability $P_{ER}''$. Then the total probability of failure including the multiplicative error in strength can be shown to be

$$P_R^{**} = (1 - P_{ER}'') \cdot P_o + P_{ER}'' \cdot \Phi(-\beta_R'') \quad (14)$$

or, alternatively, the total reliability index is given by

$$\beta_R^{**} = \Phi^{-1}(1 - P_R^{**}) \quad (15)$$

If the multiplicative error modifies both the error-free mean load effect $\bar{S}_o$ and the error-free standard derivation $\sigma(S_o)$ by the same positive factor $F_S''$, then the error-free reliability index $\beta_o$ changes to

$$\beta_S'' = (\bar{R}_o - F_S'' \cdot \bar{S}_0)/\{\sigma^2(R_o) + [F_S'' \cdot \sigma(S_o)]^2\}^{0.5} \quad (16)$$

or

$$\beta_S'' = (\theta_o - F_S'')/\{\theta_o^2 \cdot V^2(R_o) + [F_S'' \cdot V(S_0)]^2\}^{0.5} \quad (17)$$

Also, assume that the multiplicative error in load effect occurs with probability $P_{ES}''$. Then the total probability of failure including the multiplicative error in load effect can be shown to be

$$P_S^{**} = (1 - P_{ES}'') \cdot P_o + P_{ES}'' \cdot \Phi(-\beta_S'') \quad (18)$$

The positive factors $F_R'$, $F_S'$, $F_R''$ and $F_S''$ in Eqs. (4) and (5), (8) and (9), (12) and (13), and (16) and (17), respectively, could be expressed as

$$F_R' = 1 \pm m_R' \cdot V(R_o) \quad (19)$$

$$F_S' = 1 \pm m_S' \cdot V(S_o) \tag{20}$$

$$F_R'' = 1 \pm m_R'' \cdot V(R_o) \tag{21}$$

$$F_S'' = 1 \pm m_S'' \cdot V(S_o) \tag{22}$$

where $m_R'$, $m_S'$, $m_R''$ and $m_S''$ are positive numbers, and + and - correspond to values of $F_R'$, $F_S'$, $F_R''$, and $F_S''$, larger and lower than 1, respectively.

Discrete Model with Combined Errors

Consider now the case in which both errors in strength and load effect occur. These errors are supposed to be independent (i.e., cases A,B,E, and H in Table 1).

Combination of Additive Errors.- Consider first the case in which additive errors occur and modify both the error-free mean resistance $\bar{R}_o$ by a positive factor $F_R'$ and the error-free mean load effect $\bar{S}_o$ by a positive factor $F_S'$, as shown in Fig. 4. Then the error-free reliability index $\beta_o$ changes to

$$\beta_{RS}' = (F_R' \cdot \bar{R}_o - F_S' \cdot \bar{S}_o)/[\sigma^2(R_o) + \sigma^2(S_o)]^{0.5} \tag{23}$$

or

$$\beta_{RS}' = (F_R' \cdot \theta_o - F_S')/[\theta_o^2 \cdot V^2(R_o) + V^2(S_o)]^{0.5} \tag{24}$$

Also, assume that the additive errors in strength and load effect occur with probabilities $P_{ER}'$ and $P_{ES}'$, respectively. Then the total probability of failure including the additive errors is

$$P_{RS}^* = (1 - P_{ER}') \cdot (1-P_{ES}') \cdot P_o + (P_{ER}' + P_{ES}' - P_{ER}' \cdot P_{ES}') \cdot \Phi(-\beta_{RS}') \tag{25}$$

in which $(1 - P_{ER}') \cdot (1 - P_{ES}')$ is the probability of no error in strength and load effect, and $P_{ER}' + P_{ES}' - P_{ER}' \cdot P_{ES}'$ is the probability of at least one error in strength or load effect.

Alternatively, the total reliability index is

$$\beta_{RS}^* = \Phi^{-1}(1 - P_{RS}^*). \tag{26}$$

Combination of Multiplicative Errors. - Consider now the case shown in Fig. 5 in which multiplicative errors occur and modify the error-free means and standard deviations of both resistance and load effect by the positive factors $F_R''$ and $F_S''$, respectively. Then the error-free reliability index $\beta_o$ changes to

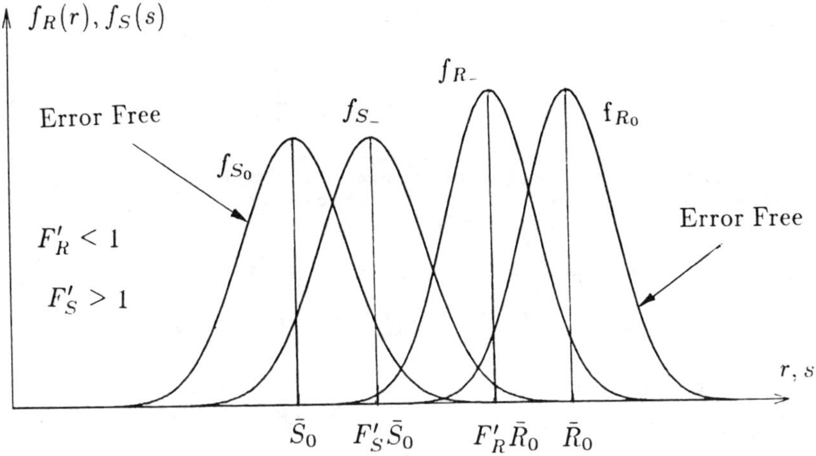


Fig. 4 - Combination of Additive Errors in Resistance and Load Effect

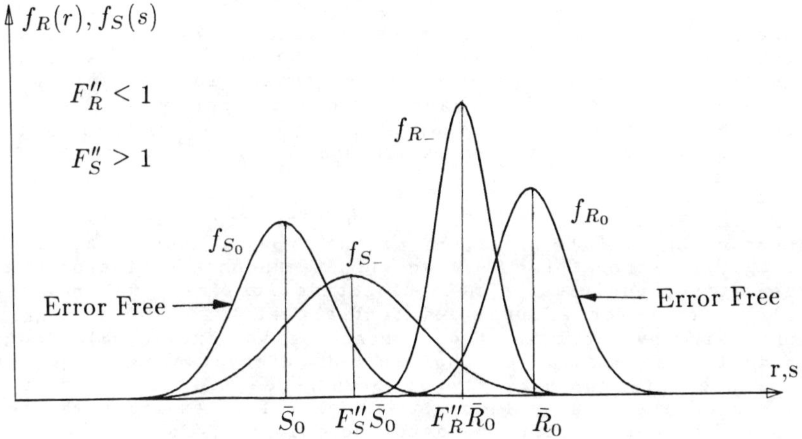


Fig. 5 - Combination of Multiplicative Errors in Resistance and Load Effect

$$\beta_{RS}'' = (F_R''.\bar{R}_o - F_S''.\bar{S}_o)/\{[F_R''.\sigma(R_o)]^2+[F_S''.\sigma(S_o]^2\}^{0.5} \quad (27)$$

or

$$\beta_{RS}'' = (F_R''.\theta_o - F_S'')/\{[F_R''.\theta_o.V(R)\}^2+[F_S''.V(S_o)]^2\}^{0.5} \quad (28)$$

Also, assume that the multiplicative errors in strength and load effect occur with probabilities $P_{ER}''$ and $P_{ES}''$, respectively. Then the total probability of failure including the multiplicative errors is

$$P_{RS}^{**} = (1-P_{ER}'').(1-P_{ES}'').P_o + (P_{ER}'' + P_{ES}'' - P_{ER}'' \cdot P_{ES}'') \cdot \Phi(-\beta_{RS}'') \quad (29)$$

in which $(1 - P_{ER}'') \cdot (1 - P_{ES}'')$ is the probability of no error in strength and load effect, and $P_{ER}'' + P_{ES}'' - (P_{ER}'' \cdot P_{ES}'')$ is the probability of at least one error in strength or load effect.

Alternatively, the total reliability index is:

$$\beta_{RS}^{**} = \Phi^{-1}(1-P_{RS}^{**}) \quad (30)$$

Numerical Examples

The data for the following numerical examples is based on values which are commonly used in structural design.

Example 1. - The purpose of this example is to investigate how the total reliability indices for individual additive and multiplicative errors in strength and load effect, $\beta_R^*$, $\beta_S^*$, $\beta_R^{**}$, and $\beta_S^{**}$, vary with the probabilities of occurrence of such errors, $P_{ER}'$, $P_{ES}'$, $P_{ER}''$, and $P_{ES}''$, respectively, for different values of the central safety factor $\theta_o$.

Tables 2 to 5 provide information for two interesting comparisons: (a) the effect of the type of individual error (e.g., in strength or in load effect, unconservative or conservative, additive or multiplicative) on the total reliability index for a specified central safety factor $\theta_o$ and a specified probability of occurrence of the individual error; and (b) the effects of the central safety factor $\theta_o$ and of the specified probability of occurrence of the individual error on the total reliability index. The results presented in Tables 2 to 5 correspond to the following data:

. coefficients of variation
. $V(R_o) = 15\%, V(S_o) = 30\%$;

. error factors for unconservative (additive or multiplicative) error in strength
$F_R' = F_R'' = 1 - 2 \cdot V(R_o) = 0.7$;

TABLE 2. - Sensitivity of Reliability Index to Changes in the Type of Individual Error for Different Values of the Probability of Error, P, and for $\theta_0$ = 2

| Probability of Error | Multiplicative Error | | | | Additive Error | | | |
|---|---|---|---|---|---|---|---|---|
| | B[a] | C | F | G | B | C | F | G |
| 0.00 | 2.36 | 2.36 | 2.36 | 2.36 | 2.36 | 2.36 | 2.36 | 2.36 |
| 0.05 | 2.37 | 2.38 | 2.15 | 2.04 | 2.38 | 2.38 | 2.11 | 2.11 |
| 0.10 | 2.39 | 2.40 | 2.01 | 1.85 | 2.40 | 2.40 | 1.95 | 1.95 |
| 0.15 | 2.41 | 2.42 | 1.90 | 1.71 | 2.42 | 2.42 | 1.83 | 1.83 |
| 0.20 | 2.43 | 2.44 | 1.81 | 1.60 | 2.44 | 2.44 | 1.73 | 1.73 |
| 0.25 | 2.45 | 2.46 | 1.74 | 1.50 | 2.46 | 2.46 | 1.64 | 1.64 |
| 0.30 | 2.48 | 2.49 | 1.67 | 1.42 | 2.49 | 2.49 | 1.57 | 1.57 |
| 0.35 | 2.50 | 2.51 | 1.61 | 1.34 | 2.51 | 2.51 | 1.50 | 1.50 |
| 0.40 | 2.53 | 2.54 | 1.55 | 1.27 | 2.54 | 2.54 | 1.44 | 1.44 |
| 0.45 | 2.55 | 2.57 | 1.50 | 1.21 | 2.57 | 2.57 | 1.39 | 1.39 |
| 0.50 | 2.58 | 2.60 | 1.45 | 1.15 | 2.60 | 2.60 | 1.33 | 1.33 |
| 0.55 | 2.62 | 2.64 | 1.41 | 1.10 | 2.64 | 2.64 | 1.29 | 1.29 |
| 0.60 | 2.65 | 2.68 | 1.37 | 1.05 | 2.68 | 2.68 | 1.24 | 1.24 |
| 0.65 | 2.69 | 2.72 | 1.33 | 1.00 | 2.72 | 2.72 | 1.20 | 1.20 |
| 0.70 | 2.73 | 2.77 | 1.29 | 0.95 | 2.77 | 2.77 | 1.16 | 1.16 |
| 0.75 | 2.78 | 2.83 | 1.25 | 0.91 | 2.83 | 2.83 | 1.12 | 1.12 |
| 0.80 | 2.83 | 2.90 | 1.22 | 0.86 | 2.89 | 2.89 | 1.08 | 1.08 |
| 0.85 | 2.90 | 2.99 | 1.18 | 0.82 | 2.98 | 2.98 | 1.04 | 1.04 |
| 0.90 | 2.98 | 3.11 | 1.15 | 0.78 | 3.09 | 3.09 | 1.01 | 1.01 |
| 0.95 | 3.09 | 3.31 | 1.12 | 0.74 | 3.27 | 3.27 | 0.98 | 0.98 |
| 1.00 | 3.25 | 4.95 | 1.09 | 0.71 | 3.77 | 3.77 | 0.94 | 0.94 |

[a]B means error combination ($R_+$, $S_0$) in Table 1
C means error combination ($R_0$, $S_+$) in Table 1
F means error combination ($R_-$, $S_0$) in Table 1
G means error combination ($R_0$, $S_-$) in Table 1

TABLE 3. - Sensitivity of Reliability Index to Changes in the Type of Individual Error for Different Values of the Probability of Error, P, and for $\theta_0$ = 3

| Probability of Error | Multiplicative Error | | | | Additive Error | | | |
|---|---|---|---|---|---|---|---|---|
| | B[a] | C | F | G | B | C | F | G |
| 0.00 | 3.70 | 3.70 | 3.70 | 3.70 | 3.70 | 3.70 | 3.70 | 3.70 |
| 0.05 | 3.71 | 3.71 | 3.36 | 3.11 | 3.71 | 3.71 | 3.05 | 3.39 |
| 0.10 | 3.72 | 3.72 | 3.21 | 2.92 | 3.72 | 3.72 | 2.85 | 3.25 |
| 0.15 | 3.74 | 3.74 | 3.11 | 2.80 | 3.74 | 3.74 | 2.72 | 3.15 |
| 0.20 | 3.75 | 3.75 | 3.03 | 2.70 | 3.75 | 3.75 | 2.63 | 3.08 |
| 0.25 | 3.77 | 3.77 | 2.97 | 2.63 | 3.77 | 3.77 | 2.55 | 3.01 |
| 0.30 | 3.78 | 3.79 | 2.91 | 2.57 | 3.79 | 3.79 | 2.49 | 2.96 |
| 0.35 | 3.80 | 3.81 | 2.87 | 2.52 | 3.81 | 3.81 | 2.44 | 2.92 |
| 0.40 | 3.82 | 3.83 | 2.83 | 2.47 | 3.83 | 3.82 | 2.39 | 2.88 |
| 0.45 | 3.84 | 3.85 | 2.79 | 2.43 | 3.85 | 3.85 | 2.35 | 2.84 |
| 0.50 | 3.86 | 3.87 | 2.76 | 2.39 | 3.87 | 3.87 | 2.31 | 2.81 |
| 0.55 | 3.88 | 3.90 | 2.73 | 2.36 | 3.90 | 3.89 | 2.27 | 2.78 |
| 0.60 | 3.91 | 3.92 | 2.70 | 2.32 | 3.92 | 3.92 | 2.24 | 2.76 |
| 0.65 | 3.94 | 3.96 | 2.67 | 2.29 | 3.96 | 3.95 | 2.21 | 2.73 |
| 0.70 | 3.97 | 3.99 | 2.65 | 2.27 | 3.99 | 3.99 | 2.18 | 2.71 |
| 0.75 | 4.00 | 4.04 | 2.63 | 2.24 | 4.04 | 4.03 | 2.15 | 2.68 |
| 0.80 | 4.05 | 4.09 | 2.60 | 2.22 | 4.09 | 4.08 | 2.12 | 2.66 |
| 0.85 | 4.10 | 4.15 | 2.58 | 2.19 | 4.15 | 4.15 | 2.10 | 2.64 |
| 0.90 | 4.17 | 4.25 | 2.56 | 2.17 | 4.25 | 4.23 | 2.08 | 2.62 |
| 0.95 | 4.26 | 4.40 | 2.55 | 2.15 | 4.40 | 4.37 | 2.06 | 2.61 |
| 1.00 | 4.41 | 5.58 | 2.53 | 2.13 | 5.36 | 4.81 | 2.03 | 2.59 |

TABLE 4. - Sensitivity of Reliability Index to Changes in the Type of Individual Error for Different Values of the Probability of Error, P, and for $\theta_0 = 4$

| Probability of Error | Multiplicative Error | | | | Additive Error | | | |
|---|---|---|---|---|---|---|---|---|
| | B[a] | C | F | G | B | C | F | G |
| 0.00 | 4.47 | 4.47 | 4.47 | 4.47 | 4.47 | 4.47 | 4.47 | 4.47 |
| 0.05 | 4.48 | 4.48 | 4.16 | 3.90 | 4.48 | 4.48 | 3.56 | 4.22 |
| 0.10 | 4.49 | 4.49 | 4.03 | 3.74 | 4.49 | 4.49 | 3.38 | 4.10 |
| 0.15 | 4.50 | 4.51 | 3.95 | 3.64 | 4.51 | 4.51 | 3.26 | 4.02 |
| 0.20 | 4.52 | 4.52 | 3.88 | 3.57 | 4.52 | 4.52 | 3.18 | 3.96 |
| 0.25 | 4.53 | 4.53 | 3.83 | 3.51 | 4.53 | 4.53 | 3.12 | 3.91 |
| 0.30 | 4.54 | 4.55 | 3.79 | 3.46 | 4.55 | 4.55 | 3.06 | 3.87 |
| 0.35 | 4.56 | 4.56 | 3.75 | 3.42 | 4.56 | 4.56 | 3.02 | 3.83 |
| 0.40 | 4.57 | 4.58 | 3.72 | 3.38 | 4.58 | 4.58 | 2.98 | 3.80 |
| 0.45 | 4.59 | 4.60 | 3.69 | 3.35 | 4.60 | 4.60 | 2.94 | 3.77 |
| 0.50 | 4.61 | 4.62 | 3.66 | 3.32 | 4.62 | 4.62 | 2.91 | 3.75 |
| 0.55 | 4.62 | 4.64 | 3.64 | 3.29 | 4.64 | 4.64 | 2.88 | 3.73 |
| 0.60 | 4.65 | 4.66 | 3.62 | 3.27 | 4.66 | 4.66 | 2.85 | 3.71 |
| 0.65 | 4.67 | 4.69 | 3.60 | 3.25 | 4.69 | 4.69 | 2.82 | 3.69 |
| 0.70 | 4.69 | 4.72 | 3.58 | 3.23 | 4.72 | 4.72 | 2.80 | 3.67 |
| 0.75 | 4.72 | 4.76 | 3.56 | 3.21 | 4.76 | 4.75 | 2.78 | 3.65 |
| 0.80 | 4.76 | 4.80 | 3.55 | 3.19 | 4.80 | 4.80 | 2.76 | 3.63 |
| 0.85 | 4.80 | 4.86 | 3.53 | 3.17 | 4.86 | 4.85 | 2.74 | 3.62 |
| 0.90 | 4.85 | 4.94 | 3.52 | 3.15 | 4.94 | 4.92 | 2.72 | 3.60 |
| 0.95 | 4.92 | 5.07 | 3.50 | 3.14 | 5.08 | 5.04 | 2.70 | 3.59 |
| 1.00 | 5.03 | 5.88 | 3.49 | 3.12 | 6.26 | 5.37 | 2.68 | 3.58 |

TABLE 5. - Sensitivity of Reliability Index to Changes in the Type of Individual Error for Different Values of the Probability of Error, P, and for $\theta_0 = 5$

| Probability of Error | Multiplicative Error | | | | Additive Error | | | |
|---|---|---|---|---|---|---|---|---|
| | B[a] | C | F | G | B | C | F | G |
| 0.00 | 4.95 | 4.95 | 4.95 | 4.95 | 4.95 | 4.95 | 4.95 | 4.95 |
| 0.05 | 4.96 | 4.96 | 4.71 | 4.48 | 4.96 | 4.96 | 3.89 | 4.76 |
| 0.10 | 4.97 | 4.97 | 4.60 | 4.34 | 4.97 | 4.97 | 3.72 | 4.66 |
| 0.15 | 4.98 | 4.98 | 4.53 | 4.26 | 4.98 | 4.98 | 3.62 | 4.59 |
| 0.20 | 4.99 | 5.00 | 4.47 | 4.19 | 5.00 | 4.99 | 3.54 | 4.54 |
| 0.25 | 5.00 | 5.01 | 4.43 | 4.14 | 5.01 | 5.01 | 3.48 | 4.49 |
| 0.30 | 5.01 | 5.02 | 4.39 | 4.10 | 5.02 | 5.02 | 3.44 | 4.46 |
| 0.35 | 5.03 | 5.03 | 4.36 | 4.07 | 5.04 | 5.03 | 3.39 | 4.43 |
| 0.40 | 5.04 | 5.05 | 4.33 | 4.04 | 5.05 | 5.05 | 3.36 | 4.40 |
| 0.45 | 5.05 | 5.07 | 4.31 | 4.01 | 5.07 | 5.06 | 3.32 | 4.38 |
| 0.50 | 5.07 | 5.08 | 4.29 | 3.98 | 5.09 | 5.08 | 3.29 | 4.36 |
| 0.55 | 5.08 | 5.10 | 4.27 | 3.96 | 5.11 | 5.10 | 3.27 | 4.34 |
| 0.60 | 5.10 | 5.13 | 4.25 | 3.94 | 5.13 | 5.12 | 3.24 | 4.32 |
| 0.65 | 5.12 | 5.15 | 4.23 | 3.92 | 5.15 | 5.15 | 3.22 | 4.30 |
| 0.70 | 5.14 | 5.18 | 4.21 | 3.90 | 5.18 | 5.17 | 3.20 | 4.29 |
| 0.75 | 5.17 | 5.21 | 4.20 | 3.89 | 5.22 | 5.21 | 3.18 | 4.27 |
| 0.80 | 5.20 | 5.25 | 4.18 | 3.87 | 5.26 | 5.24 | 3.16 | 4.26 |
| 0.85 | 5.23 | 5.31 | 4.17 | 3.86 | 5.31 | 5.29 | 3.14 | 4.24 |
| 0.90 | 5.27 | 5.38 | 4.16 | 3.84 | 5.38 | 5.36 | 3.13 | 4.23 |
| 0.95 | 5.32 | 5.50 | 4.15 | 3.83 | 5.51 | 5.46 | 3.11 | 4.22 |
| 1.00 | 5.39 | 6.06 | 4.13 | 3.82 | 6.86 | 5.69 | 3.09 | 4.21 |

- error factors for conservative (additive or multiplicative) error in strength
  $F_R' = F_R'' = 1 + 2 \cdot V(R_o) = 1.3$;
- error factors for unconservative (additive or multiplicative) error in load effect
  $F_S' = F_S'' = 1 + 2 \cdot V(S_o) = 1.6$;
- error factors for conservative (additive or multiplicative) error in load effect
  $F_S' = F_S'' = 1 - 2 \cdot V(S_o) = 0.4$

It is also assumed that the probabilities of occurrence of all the errors (i.e., additive or multiplicative, in strength or in load effect) are all the same:

$$P_{ER}' = P_{ES}' = P_{ER}'' = P_{ES}'' = P$$

It appears from the results presented in Tables 2 to 5 that:

(a) the additive unconservative errors in strength have more effect on the total reliability index that the multiplicative errors. This sensitivity of the global reliability index to the type of unconservative error in strength increases with $\theta_o$ (see columns F in Tables 2-5):

(b) the global reliability index is almost insensitive to the type of conservative error in strength (see columns $_B$ in Tables 2-5); this is especially true if the probability of occurrence of these errors is not large, (say, $P < 0.60$);

(c) the multiplicative unconservative errors in load effect have more effect on the total reliability index than the additive errors (see columns G in Tables 2-5);

(d) the global reliability index is almost insensitive to the type of conservative error in load effect (see columns C in Tables 2-5). This is true for approximately all the range of possible values of the probability of occurrence of these errors (say, $P < 0.90$);

(e) it is generally more dangerous to have multiplicative (additive) unconservative errors in load effect (strength) than in strength (load effect); and

(f) it is generally more advantageous to have additive (multiplicative) conservative errors in strength (load effect) than in load effect (strength).

Example 2. - Let us also consider the case of combination of errors. The purpose of this second example is to investigate how the total reliability indices for both individual and combined errors change with the probabilities of occurrence of such errors.

Tables 6 and 7 compare global reliability indices, for additive and multiplicative errors, respectively. For each one of the eight cases of combination of errors listed in Table 1, Tables 7 and 8 present the global reliability indices as a function of the probability P of occurrence of such errors. The results presented in these tables correspond to the following data:

- central safety factor
  $\theta_o = \bar{R}_o/\bar{S}_o = 3$;
- coefficients of variation
  $V(R_o) = 15\%$, $V(S_o) = 25\%$;
- error factors for unconservative (additive or multiplicative) error in strength
  $F_R' = F_R'' = 1 - 2 \cdot V(R_o) = 0.7$;
- error factors for conservative (additive or multiplicative) error in strength
  $F_R' = F_R'' = 1 + 2 \cdot V(R_o) = 1.3$;
- error factors for unconservative (additive or multiplicative) error in load effect
  $F_S' = F_S'' = 1 + 2 \cdot V(S_o) = 1.5$;
- error factors for conservative (additive or multiplicative) error in load effect
  $F_S' = F_S'' = 1 - 2 \cdot V(S_o) = 0.5$.

It is also assumed that the probabilities of occurrence of all the errors are the same:

$$P_{ER}' = P_{ES}' = P_{ER}'' = P_{ES}'' = P$$

As expected, because of the different ways to combine errors, the various cases considered do not yield identical results, with the exception of the case where P = 0. It is obvious that the larger (smaller) total reliability indices are obtained in the case where conservative (unconservative) errors in strength are combined with conservative (unconservative) errors in load effect [see columns A and H in Tables 6 and 7]. It is interesting to observe that in the case where conservative additive (multiplicative) errors in strength are combined with unconservative additive (multiplicative) errors in load effect the total reliability index increases (decreases) with increasing the probability P of occurrence of such errors [see columns D in Tables 6 and 7]. It is also interesting to observe that combining unconservative additive (multiplicative) errors in strength with conservative additive (multiplicative) errors in load effect shows a reduction (an increase) in the total reliability index [see columns E in Tables 6 and 7]; this reduction (increase) is larger when the probability P is higher.

TABLE 6. - **Sensitivity of Reliability Index to Changes in Both Individual and Combined Additive Errors for Different Values of the Probability of Error, P, and for $\theta_0$ = 3**

| Probability of Error | Individual Error | | | | Combined Errors | | | |
|---|---|---|---|---|---|---|---|---|
| | B[a] | C | F | G | A | D | E | H |
| 0.00 | 3.89 | 3.89 | 3.89 | 3.89 | 3.89 | 3.89 | 3.89 | 3.89 |
| 0.05 | 3.90 | 3.90 | 3.13 | 3.64 | 3.91 | 3.91 | 3.64 | 2.26 |
| 0.10 | 3.91 | 3.91 | 2.93 | 3.51 | 3.94 | 3.93 | 3.51 | 1.99 |
| 0.15 | 3.92 | 3.92 | 2.81 | 3.42 | 3.96 | 3.96 | 3.43 | 1.83 |
| 0.20 | 3.94 | 3.94 | 2.72 | 3.35 | 3.99 | 3.99 | 3.37 | 1.71 |
| 0.25 | 3.95 | 3.95 | 2.64 | 3.30 | 4.02 | 4.02 | 3.33 | 1.61 |
| 0.30 | 3.97 | 3.97 | 2.58 | 3.25 | 4.06 | 4.05 | 3.29 | 1.54 |
| 0.35 | 3.99 | 3.99 | 2.53 | 3.21 | 4.09 | 4.08 | 3.26 | 1.47 |
| 0.40 | 4.01 | 4.01 | 2.48 | 3.18 | 4.13 | 4.11 | 3.23 | 1.42 |
| 0.45 | 4.03 | 4.03 | 2.44 | 3.14 | 4.17 | 4.15 | 3.21 | 1.37 |
| 0.50 | 4.05 | 4.05 | 2.40 | 3.12 | 4.21 | 4.19 | 3.19 | 1.33 |
| 0.55 | 4.07 | 4.07 | 2.37 | 3.09 | 4.26 | 4.23 | 3.17 | 1.30 |
| 0.60 | 4.10 | 4.10 | 2.33 | 3.06 | 4.31 | 4.28 | 3.16 | 1.27 |
| 0.65 | 4.13 | 4.13 | 2.30 | 3.04 | 4.37 | 4.32 | 3.14 | 1.24 |
| 0.70 | 4.17 | 4.16 | 2.28 | 3.02 | 4.44 | 4.38 | 3.13 | 1.22 |
| 0.75 | 4.21 | 4.20 | 2.25 | 3.00 | 4.51 | 4.43 | 3.13 | 1.20 |
| 0.80 | 4.26 | 4.25 | 2.22 | 2.98 | 4.61 | 4.49 | 3.12 | 1.19 |
| 0.85 | 4.32 | 4.31 | 2.20 | 2.96 | 4.73 | 4.55 | 3.11 | 1.18 |
| 0.90 | 4.41 | 4.39 | 2.18 | 2.95 | 4.89 | 4.61 | 3.11 | 1.17 |
| 0.95 | 4.56 | 4.52 | 2.16 | 2.93 | 5.15 | 4.65 | 3.11 | 1.17 |
| 1.00 | 5.63 | 4.86 | 2.14 | 2.91 | 6.60 | 4.66 | 3.11 | 1.17 |

[a]A means error combination ($R_+$, $S_+$) in Table 1
B means error combination ($R_+$, $S_0$) in Table 1
C means error combination ($R_0$, $S_+$) in Table 1
D means error combination ($R_+$, $S_-$) in Table 1
E means error combination ($R_-$, $S_+$) in Table 1
F means error combination ($R_-$, $S_0$) in Table 1
G means error combination ($R_0$, $S_-$) in Table 1
H means error combination ($R_-$, $S_-$) in Table 1

TABLE 7. - **Sensitivity of Reliability Index to Changes in Both Individual and Combined Multiplicative Errors for Different Values of the Probability of Error, P, and for $\theta_0$ = 3**

| Probability of Error | Individual Error | | | | Combined Errors | | | |
|---|---|---|---|---|---|---|---|---|
| | B[a] | C | F | G | A | D | E | H |
| 0.00 | 3.89 | 3.89 | 3.89 | 3.89 | 3.89 | 3.89 | 3.89 | 3.89 |
| 0.05 | 3.90 | 3.90 | 3.53 | 3.42 | 3.91 | 3.80 | 3.91 | 2.30 |
| 0.10 | 3.91 | 3.91 | 3.38 | 3.25 | 3.94 | 3.73 | 3.93 | 2.03 |
| 0.15 | 3.92 | 3.92 | 3.28 | 3.15 | 3.96 | 3.69 | 3.96 | 1.87 |
| 0.20 | 3.94 | 3.94 | 3.21 | 3.07 | 3.99 | 3.65 | 3.99 | 1.75 |
| 0.25 | 3.95 | 3.95 | 3.15 | 3.00 | 4.02 | 3.62 | 4.02 | 1.66 |
| 0.30 | 3.97 | 3.97 | 3.10 | 2.95 | 4.06 | 3.59 | 4.05 | 1.59 |
| 0.35 | 3.98 | 3.99 | 3.06 | 2.90 | 4.09 | 3.57 | 4.08 | 1.52 |
| 0.40 | 4.00 | 4.01 | 3.02 | 2.86 | 4.13 | 3.55 | 4.12 | 1.47 |
| 0.45 | 4.02 | 4.03 | 2.98 | 2.82 | 4.17 | 3.53 | 4.15 | 1.43 |
| 0.50 | 4.04 | 4.05 | 2.95 | 2.79 | 4.21 | 3.51 | 4.19 | 1.39 |
| 0.55 | 4.06 | 4.07 | 2.92 | 2.76 | 4.26 | 3.50 | 4.24 | 1.35 |
| 0.60 | 4.09 | 4.10 | 2.90 | 2.73 | 4.31 | 3.49 | 4.28 | 1.32 |
| 0.65 | 4.11 | 4.13 | 2.87 | 2.71 | 4.37 | 3.48 | 4.33 | 1.30 |
| 0.70 | 4.14 | 4.17 | 2.85 | 2.68 | 4.43 | 3.47 | 4.39 | 1.28 |
| 0.75 | 4.18 | 4.21 | 2.83 | 2.66 | 4.51 | 3.47 | 4.45 | 1.26 |
| 0.80 | 4.22 | 4.26 | 2.81 | 2.64 | 4.61 | 3.46 | 4.51 | 1.25 |
| 0.85 | 4.27 | 4.32 | 2.79 | 2.62 | 4.72 | 3.46 | 4.58 | 1.24 |
| 0.90 | 4.33 | 4.41 | 2.77 | 2.60 | 4.88 | 3.46 | 4.65 | 1.23 |
| 0.95 | 4.42 | 4.56 | 2.75 | 2.58 | 5.14 | 3.45 | 4.70 | 1.23 |
| 1.00 | 4.56 | 5.35 | 2.74 | 2.56 | 5.68 | 3.45 | 4.72 | 1.23 |

## Conclusion

It has been shown that including human errors in probabilistic models for structural risk assessment is a practical possibility. The minimum information required is the error factor of the resistance and/or load effect plus the type of error (i.e., additive or multiplicative). Simplified probabilistic models of combining human errors such as the models presented herein offer the potential for integrating errors in structural risk assessment. Further research is needed to establish more comprehensive models to consistently deal with insufficient knowledge and data on human errors and to implement such models in the development of practical design criteria and code provisions.

## Acknowledgement

This work was supported, in part, by the Natural Hazards Research and Information Center, Boulder, Colorado. Thanks are due to Rachid Nakib, graduate student, for performing numerical calculations. Patricia Wathen carefully typed this manuscript.

## References

Allen, D.E., Criteria for Design Safety Factors and Quality Assurance Expenditures, Proceedings, 3rd International Conference on Structural Safety and Reliability, Trondheim, Norway, June 23-25, 1981, pp. 667-678.

Allen, D.E., Structural Failures Due to Human Errors; What Research to Do?, in Risk, Structural Engineering and Human Error (ed. M. Grigoriu), University of Waterloo Press, Waterloo, Ontario, Canada, 1984, pp. 127-137.

Blockley, D.I., Reliability Theory - Incorporating Gross Errors, Proceedings, 3rd International Conference on Structural Safety and Reliability, Trondheim, Norway, June 23-25, 1981, pp. 259-282.

Blockley, D.I., The Nature of Structural Design and Safety, Ellis Wordwood Limited, England, 1980.

Frangopol, D.M., Nakib, R., and Trautner, J., Influence of Human Errors on Structural Safety, Report, Institute of Behavioral Science, Natural Hazards Research and Information Center, Boulder, Colorado, August, 1985.

Frangopol, D.M., and Trautner, J.J., Predicting the Sensitivity of Structural Safety to Errors, Proceedings, 19th Midwestern Mechanics Conference, Columbus, Ohio, Sept. 9-11, 1985, pp. 516-517.

Grigoriu, M., (editor), Risk, Structural Engineering and Human Error, University of Waterloo Press, Waterloo,

Ontario, Canada, 1984.

Knoll, F., Human Error in the Building Process - A Research Proposal, IABSE Journal, J-11/82, Nov., 1982.

Knoll, F., Modelling Gross Errors: Some Practical Considerations, in Risk; Structural Engineering and Human Error, University of Waterloo Press, Waterloo, Ontario, Canada, 1984.

Lind, N.C., Models of Human Error in Structural Reliability, Structural Safety, 1, 1983, pp. 167-175.

Lind, N.C., Management of Gross Error, Proceedings, 4th International Conference on Applications of Statistics and Probability in Soil and Structural Engineering, Florence, Italy, June 13-17, 1983, pp. 669-681.

Matousek, M., A system for a Detailed Analysis of Structural Failures, Proceedings, 3rd International Conference on Structural Safety and Reliability, Trondheim, Norway, June 23-25, 1981, pp. 535-544.

Melchers, R.R., and Harrington, M.V., Structural Safety as Affected by Human Error, Proceedings, 4th International Conference on Applications of Statistics and Probability in Soil and Structural Engineering, Florence, Italy, June 13-17, 1983, pp. 683-694.

Nowak, A.S., Effect of Human Error on Structural Safety, ACI Journal, 1979, pp. 959-972.

Nowak, A.S. and Carr, R.I., Sensitivity Analysis for Human Error, Journal of Structural Engineering, Vol. 111, No. 8, Aug. 1985, pp. 1734-1746.

Rackwitz, R., Note on the Treatment of Errors in Structural Reliability, Technische Universitat Munchen, Ingenierbau, Heft 21, 1977, pp. 23-35.

Rosenblueth, E., On the Processing of Doubtful Information, Parts 1 and 2, in Risk, Structural Engineering and Human Error (ed M. Grigoriu), University of Waterloo Press, Waterloo, Ontario, Canada, 1984, pp. 33-88.

Schneider, J., Organization and Management of Structural Safety during Design, Construction and Operation of Structures, Proceedings, 3rd International Conference on Structural Safety and Reliability, Trondheim, Norway, June 23-25, 1981, pp. 467-482.

## EFFECTS OF HUMAN ERROR IN ANALYSIS

Mircea D. Grigoriu, M. ASCE*

### Abstract

Examples from ocean engineering and nonlinear dynamics are used to illustrate human errors in analysis and evaluate effects of these errors on structural reliability. The study investigates errors related to unsatisfactory approximations in analysis and uncertainties in various structural and loading parameters. It is shown that human error in analysis can result in unsafe designs.

### Introduction

Human error in analysis, design, construction, and use is the cause of a significant number of catastrophic structural failures. Human error cannot be eliminated from the design process. However, the rate of occurrence of this error can be reduced significantly by measures of quality assurance. The measures can include checking of calculations and assumptions in analysis, inspection of construction site, and other procedures.

This paper uses examples from ocean engineering and nonlinear dynamics to illustrate common (human) errors in analysis and evaluate effects of these errors on the predicted structural response. The study examines errors caused by unsatisfactory approximations and inaccurate representations of the uncertainty in various structural parameters. It is shown that human error in analysis can result in unsafe designs.

### Approximations In Analysis

Approximations are common to engineering analyses. The modeling of structures and loads, the determination of internal forces and deformations, and the sizing of structural members require approximations due to either the complexity of the task or the limited available information. Two examples are used to illustrate effects of approximations that currently are applied in engineering to estimate extreme wave forces and characterize the response of nonlinear dynamic systems.

---

*Associate Professor of Structural Engineering
Cornell University, Ithaca, NY 14853

## Wave Forces

Let $Y^*(t) = y_o + Y(t)$ be the wave particle velocity at time t. Assume that Y(t) is a twice-differentiable, stationary Gaussian process with mean zero and variance $\sigma_Y^2$. The constant $y_o$ denotes the fluid current. According to Morison's equation (Grigoriu 1984, Grigoriu and Alibe 1986), wave forces are proportional to

$$X(t) = X_1(t) + X_2(t) \tag{1}$$

in which the processes

$$X_1(t) = (y_o + Y(t))\ |y_o + Y(t)| \tag{2}$$

and

$$X_2(t) = a\ \dot{Y}(t) \tag{3}$$

denote drag and inertia forces and $\dot{Y}(t) = dY(t)/dt$ = a zero-mean stationary Gaussian process with variance $\sigma_{\dot{Y}}^2$. It is assumed for simplicity that drag forces are dominant so that $X(t) \simeq X_1(t)$.

Denote by $X_{1,\tau}$ the maximum of $X_1(t)$ during a storm of a duration $\tau$. Figure 1 from (Grigoriu 1984) shows with solid lines the variation of the first four moments of $X_{1,\tau}$ with

$$\upsilon_o \tau = \frac{1}{2\pi} \frac{\sigma_{\dot{Y}}}{\sigma_Y} \quad \text{for } \alpha = \sigma_Y / y_o = 0.5.$$

Results are based on an asymptotic approximation of the distribution of $X_{1,\tau}$ depending on the mean upcrossing rate function of $X_1(t)$. The dotted lines in the figure represent approximate moments of $X_{1,\tau}$ derived under the hypothesis that the drag force $X_1(t)$ is a Gaussian process. From Fig. 1, the Gaussian hypothesis can underestimate significantly both the mean and the variance of the peak wave force $X_{1,\tau}$. Therefore, it should not be used to estimate design wave forces.

## Equivalent Linearization

"Equivalent" linear systems are used frequently in engineering to approximate the behavior of nonlinear dynamic systems. The parameters of the equivalent linear systems can be obtained from the minimization of an objective function (Nayfeh and Mook 1979). Direct observations of the nonlinear equation of motion can also be considered to develop linear approximations (Huntley and Johnson 1983).

Consider, e.g., the Van der Pol equation

$$\ddot{x}(t) + 2\ \zeta(1-x^2(t))\ \dot{x}(t) + x(t) = 0 \tag{4}$$

with the initial conditions $x(0) = x_o$ and $\dot{x}(0) = 0$. The

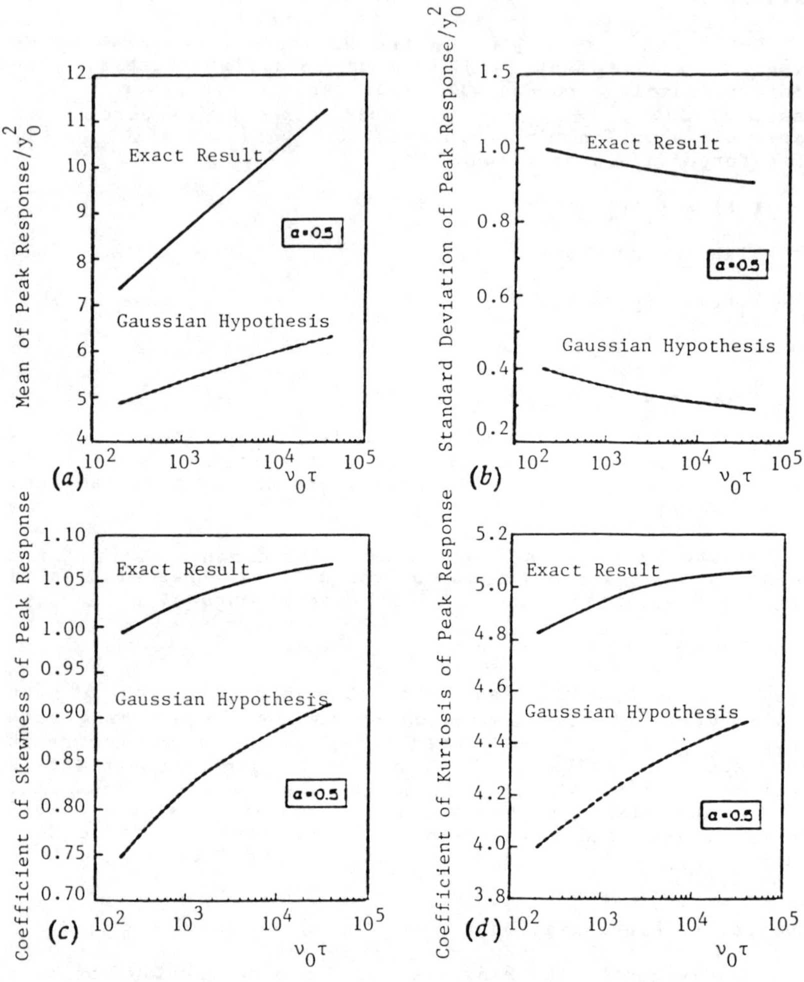


Fig. 1 Exact and Approximate Moments of Peak Drag Force.

equation can model electric circuits or various mechanical systems. For small amplitudes $1 - x^2(t) \simeq 1.0$ so that Eq. 4 can be approximated by a linear differential equation with the solution

$$x(t) = x_o e^{-\zeta t} \left[\cos(\sqrt{1-\zeta^2}\, t) + \frac{\zeta}{\sqrt{1-\zeta^2}} \sin(\sqrt{1-\zeta^2}\, t)\right] \quad (5)$$

From Eq. 5, $x(t) \to 0$ as $t \to \infty$ for any value of $x_o$. On the other hand, the solution of Eq. 4 becomes unbounded as $t \to \infty$ if $x_o > 2$. Fig. 2 from (Huntley and Johnson 1983) shows exact solutions of the Van der Pol equation for two initial conditions, $x_o = 1$ and $x_o = 2.02$. The use of linear approximation for initial states $x_o > 2$ constitutes a gross human error. If this error occurs, systems that will fail with certainty, due to the unlimited increase of x(t) in time when $x_o > 2$, are thought to be safe.

The equivalent linearization method has been applied with satisfactory result, e.g., in offshore engineering to estimate the mean and variance of the response of marine structures to wave forces. However, the method provides unconservative estimates of the peak response primarily due to the intrinsic assumption in this method that the structural response follows a Gaussian distribution. Fig. 3 shows averages of the one-hour peak response of an offshore platform modeled as a simple oscillator and subject to wave forces following the interactive form of Morison's equation. The analysis is based on simulation and statistical linearization (Grigoriu and Alibe 1986). Note that the linearization method is generally unconservative and is less satisfactory for large current ratios $y_o/\sigma_Y$. Nevertheless, it has been applied extensively in structural dynamics to estimate the peak response. This human error in analysis seems to be caused by a lack of understanding of the limitations of the linearization method.

## Uncertainty In Analysis

The analysis of dynamic systems is based on mathematical models consisting of differential equations of motion, initial conditions, and forcing functions. The functional form of the models and the values of the parameters of these models are usually uncertain. The uncertainty can be caused by an insufficient understanding of system behavior (human error) and/or a limited information on various system characteristics. Two examples are used to evaluate effects of the uncertainty in modeling. Subjective probabilities are applied to represent the uncertainty in the parameters of a particular system. Fuzzy sets could also be employed for this purpose.

## Parametrically Excited Systems

Consider a simply supported column subject to a fluctuating axial load. The transversal oscillations of the col-

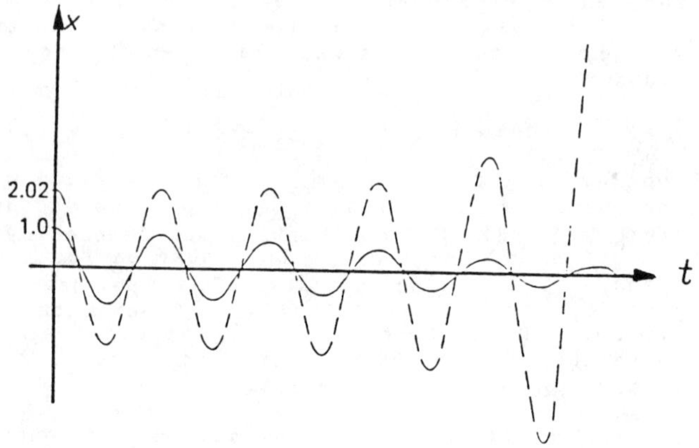


Fig. Solution of the Van der Pol Equation.

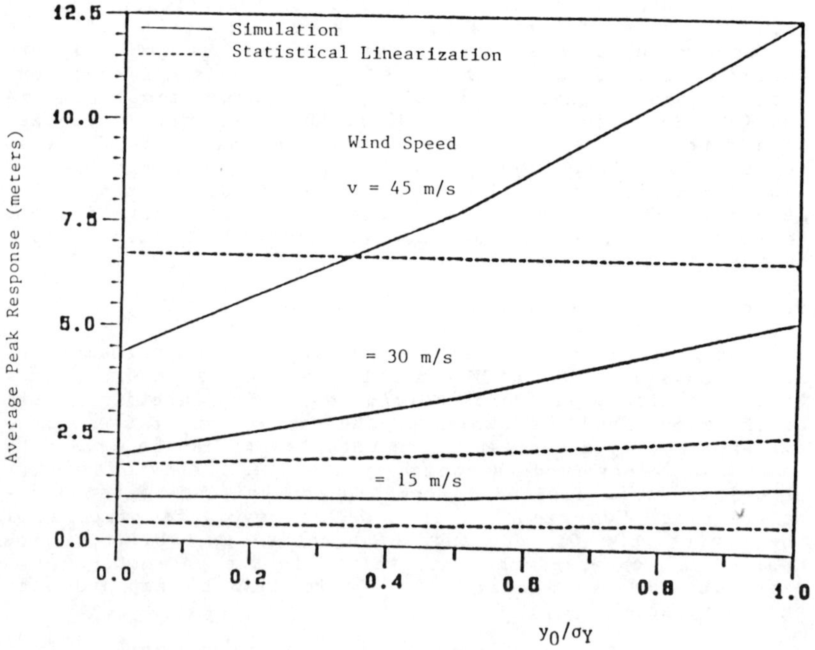


Fig. 3 Average Peak Surge Response of Offshore Platforms.

umn can be characterized approximately by a function x(t) satisfying the Mathieu-Hill equation (Nayfeh and Mook 1979)

$$\ddot{x}(t) + 2\zeta\,\dot{x}(t) + \omega^2\,(1-2\mu\cos\upsilon t)\,x(t) = 0 \qquad (6)$$

in which $\upsilon = 2\pi/T$ and $\mu$ = frequency and the intensity parameters of the axial load, $\omega$ = the natural frequency of the column in the first mode, and $\zeta$ = the corresponding modal damping ratio.

It can be shown that the oscillations of the column become asymptotically unstable (i.e., $|x(t)| \to \infty$ as $t \to \infty$) if $\xi = \upsilon/2\omega$ is in the range $(\xi_1(\rho), \xi_2(\rho))$, in which $\xi_1(\rho) = \sqrt{1-\sqrt{\mu^2-\rho^2}}$ $\xi_2(\rho) = \sqrt{1+\sqrt{\mu^2-\rho^2}}$, $\rho=2\zeta/\omega$ and $\rho<\mu$. Fig. 4 shows the primary region of asymptotic instability in the $(\rho, \xi)$ space for $\mu = 0.2$. The probability that the oscillations become unstable as $t\to\infty$ can be obtained for any value of $\mu$ from

$$P_F(\mu) = \int_0^{\mu} d\rho\, f(\rho) \int_{\xi_1(\rho)}^{\xi_2(\rho)} d\xi\, f(\xi) \qquad (7)$$

in which $f(\rho)$ and $f(\xi)$ = (subjective) probability density functions of $\rho$ and $\xi$.

Table 1 gives probabilities $P_F(\mu)$ in Eq. 7 for $\mu = 0.2$,

$$f(\rho) = \begin{cases} 175.0\,(2.5\rho)^4\,(1-2.5\rho)^4, & \text{case 1} \\ 33.84\,(2.5\rho)^{0.5}\,(1-2.5\rho)^4, & \text{case 2} \\ 33.84\,(2.5\rho)^4\,(1-2.5\rho)^{0.5}, & \text{case 3} \end{cases}$$

$$f(\xi) = \frac{1}{\sqrt{2\pi}\,\sigma} \exp\left[-\frac{1}{2}\left(\frac{\xi-m}{\sigma}\right)^2\right] \qquad (8)$$

m = 1, and $\sigma$ = 0.1 and 0.2. Results show that the probability of an asymptotically unstable motion depends strongly on the model of $\rho = 2\zeta/\omega$. Overconservative or unconservative design can result when the damping parameter $\zeta$ is overestimated (case 2) or underestimated (case 3) relative to the reference situation (case 1).

Table 1. Motion Instability for the Mathieu-Hill Equation

| Case | Probability of Asymptotic Instability | |
|---|---|---|
| | m = 1 ; $\sigma$ = 0.1 | m = 1 ; $\sigma$ = 0.2 |
| 1 | $2.45 \times 10^{-2}$ | $1.30 \times 10^{-2}$ |
| 2 | $5.47 \times 10^{-1}$ | $3.11 \times 10^{-1}$ |
| 3 | $2.46 \times 10^{-2}$ | $1.28 \times 10^{-2}$ |

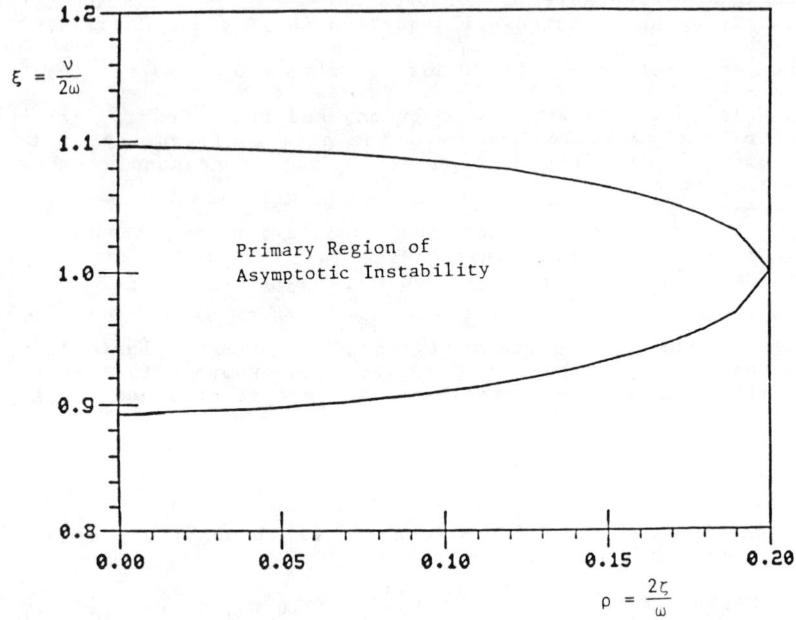


Fig. 4 Primary Region of Asymptotic Instability for the Mathieu-Hill Equation.

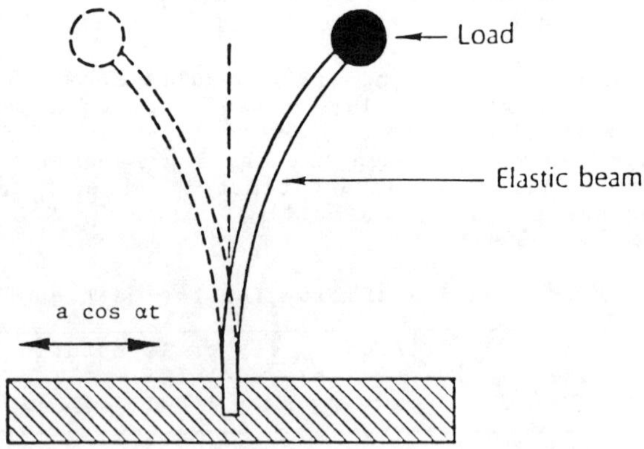


Fig. 5 Oscillating Buckled Beam.

Chaotic Motion

Consider the vertical cantilever in Fig. 5 supporting a gravity load at its free end. The cantilever is subjected to the harmonic excitation a cos $\alpha$t. The oscillations of this system can be modeled approximately by the Duffing equation (Holmes and Moon 1983) .

$$\ddot{x}(t) + \zeta\dot{x}(t) - \frac{1}{2}(1 - x^2(t))\,x(t) = a\cos\alpha t \qquad (9)$$

in which $\zeta$ = a damping parameter and x(t) = the displacement of the cantilever's tip relative to its base.

The response x(t) is periodic for small excitations but becomes chaotic (i.e., the cantilever has a random-like unpredictable motion) as the amplitude a of the forcing function increases. A necessary condition for chaos is (Holmes and Moon 1983)

$$a > \frac{\sqrt{2}\,\zeta}{3\pi\alpha}\cosh\left(\frac{\pi\alpha}{\sqrt{2}}\right) \qquad (10)$$

Fig. 6 illustrates this condition in the ($\alpha$, $\zeta$) space for a = 0.05 and 0.10.

Table 2 gives probabilities

$$P_F(a) = \int d\alpha\, f(\alpha) \int_0^{\gamma^*(\alpha)} d\zeta\, f(\zeta) \qquad (11)$$

of chaotic motion for forcing functions of specified amplitudes a. It is assumed that

$$f(\alpha) = \frac{1}{\sqrt{2\pi}\,\sigma}\exp\left[-\frac{1}{2}\left(\frac{\alpha - m}{\sigma}\right)^2\right]$$

$$f(\zeta) = \frac{\zeta^{k-1}\lambda^k e^{-\lambda\zeta}}{\Gamma(k)} \qquad (12)$$

m = 0.35; 0.45; 0.55, $\sigma$ = 0.1, k = 11, and $\lambda$ = 75. The upper limit $\gamma^*(\alpha)$ in Eq. 11 is

$$\gamma^*(\alpha) = \frac{3\pi}{\sqrt{2}}\,\frac{a\alpha}{\cosh(\pi\alpha/\sqrt{2})}$$

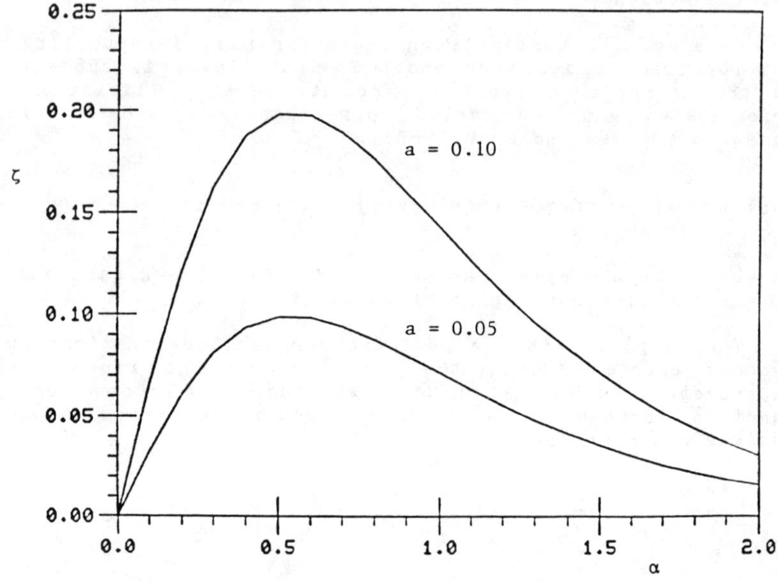


Fig. 6 Region of Chaotic Motion for the Beam in Fig. 5.

Table 2. Chaotic Motion

| a | Probability of Chaotic Motion | | |
|---|---|---|---|
| | m = 0.35 | m = 0.45 | m = 0.55 |
| 1 | $7.21 \times 10^{-2}$ | $1.06 \times 10^{-1}$ | $1.20 \times 10^{-1}$ |
| 2 | $6.96 \times 10^{-2}$ | $8.19 \times 10^{-1}$ | $8.54 \times 10^{-1}$ |

and follows from the condition in Eq. 12. Results show that the relatively small errors in the representations of the frequency of the forcing function can result in significant differences in the probability of occurrence of chaotic motion.

## Conclusion

Effects of human error in analysis have been evaluated for several engineering problems of practical interest. It has been shown that unsafe structures can result when the analysis is based mistakenly on some approximations and/or the uncertainty in various structural parameters is modeled inaccurately. The effects of such human errors in analysis on the estimated response can be significant for nonlinear dynamic systems.

## References

Grigoriu, M., "Extremes of Wave Forces," Journal of Engineering Mechanics, ASCE, Vol. 110, No. 12, December, 1984, pp. 1731-1742.

Grigoriu, M., and Alibe, B., "Response of Offshore Structures to Random Waves," Journal of Engineering Mechanics, ASCE, August, 1986.

Holmes, P. J., and Moon, F. E., "Strange Attractors and Chaos in Nonlinear Mechanics," Journal of Applied Mechanics, Vol. 50, December, 1983, pp. 1021-1032.

Huntley, I. D., and Johnson, R. M., Linear and Nonlinear Differential Equations, John Wiley & Sons, N.Y., 1983.

Nayfeh, A. H., and Mook, D. T., Nonlinear Oscillations, John Wiley & Sons, N.Y., 1978.

# SENSITIVITY ANALYSIS FOR HUMAN ERRORS

Abdulrahim M. Arafah*, and Andrzej S. Nowak, M. ASCE**

Abstract

An important objective of error control efforts is to reduce the risk resulting from variation of structural parameters beyond tolerance limits. Human errors are the major cause of such variations. In this paper, an approach is proposed to improve the effectiveness of the error control based on the sensitivity and reliability analysis. Sensitive parameters are identified and the control efforts can be allocated accordingly. Sensitivity analysis is demonstrated on two structures: a prestressed concrete girder bridge system and a steel frame structure. The developed sensitivity functions point to the parameters that deserve more quality control attention during design and construction.

## Introduction

Human error is defined as a departure from acceptable practice (Nowak and Carr 1985). It is considered as the major cause of structural failures (Matousek 1977). Structural errors are defined as the out-of-tolerance variations of basic strength parameters. Such variation can be expressed in terms of the percentage of departure from the error-free value of the parameters. The structural reliability is measured in terms of the reliability index, $\beta$ . The error-free value of $\beta$ can be calculated using the available procedures. However, only very few approaches have been proposed as a means to calculate, $\beta$, including uncertainty due to errors (Lind 1983, Rackwitz 1983).

In the last decade, the concept of modeling human errors has been considered in civil engineering (Lind 1983, Nowak and Carr 1985) and in other engineering fields (Dhillon 1983, Ravindra 1985, Swain and Guttmann 1983). Several approaches have been developed to integrate human errors into structural reliability models. The main objectives of these approaches are to describe human error characteristics such as error type, magnitude, consequences, detection and control, and to improve the existing quality control methods or to develop new control strategies.

---

*Research Assistant, Dept. of Civil Engineering, University of Michigan, Ann Arbor, MI 48109

**Associate Professor of Civil Engineering, University of Michigan, Ann Arbor, MI 48109

There are two requirements that need to be met in order to develop an effective error control process. The first one is to establish a relationship between the conditions that may lead to errors and the rate of error occurrence. This requires the development of an error data bank. The data can be stored in the form of fault trees, including a large set of error scenarios that may lead to an increased failure probability. The conditions associated with each scenario should be recorded. The second requirement is to develop a measure for error consequences. The effect of errors can be measured in terms of the increase in the expected total cost due to increased structural uncertainty. This allows for identification of the most consequential error scenarios and accordingly, the optimization of control efforts.

In this study, the effect of human error is modeled as a chain of cause and effect events in a given process. A human error causes a structural inadequacy which causes a reduction in structural reliability, and this in turn leads to an increase in the expected total cost of the process. A process can include macro-tasks such as structural planning, design, drafting, and construction, or micro-tasks such as design of foundations, design of columns, slabs or connections.

The proposed model requires the development of the following three relationships:

(1) The effect of error occurrence on the basic structural parameters.
(2) The relationship between these parameters and the structural reliability.
(3) The relationship between structural reliability and the expected cost of the process.

These relationships are described in by Arafah (1986). In this paper, the relation between the basic parameters and structural reliability is investigated using sensitivity analysis.

## Sensitivity Analysis

Sensitivity analysis can be used to establish the relationship between the basic parameters and structural reliability (Nowak and Carr 1985). The objective of such an analysis is to identify the most sensitive structural parameters. This in turn allows for the most effective allocation of control efforts. way. In particular, this will improve the checking and inspection processes during structural design and construction. Sensitivity analysis includes the following steps:

1. Formulate the required limit state function, for example, ultimate bending, shear, buckling, deflection,

or fatigue.

2. Identify basic parameters such as area of steel, strength of concrete, and section dimensions.
3. Evaluate the error-free reliability (in terms of the reliability index $\beta$ ).
4. Reformulate the limit state function to include an error in one of the design parameters. The magnitude of the error can be expressed as a percentage of the correct value of the parameter.
5. Reevaluate the structural reliability to include the error.
6. Increase the error percentage and repeat steps 4 and 5.
7. Repeat steps 4 through 6 for other basic parameters.
8. Plot the resulting sensitivity functions (reliability index vs. error percentage).

Sensitivity analysis can be extended to investigate the impact of errors from the structural system point of view. Such analysis may include the following steps:

1. Identify the structural modes of failure.
2. Formulate the limit state function for each mode of failure.
3. Calculate the probability of failure for each mode.
4. Define the failure critera for the structural system.
5. Evaluate the upper and lower bounds of the system probability of failure.
6. Evaluate the sensitivity of these bounds with respect to errors in related structural parameters.

Sensitivity analysis of a structural system is demonstrated using two numerical examples. The first example is a simply supported girder bridge system, and the second one is a steel frame structure.

Sensitivity Analysis For A Girder Bridge System

The bridge system consists of five prestressed concrete girders and a cast-in-place concrete slab as shown in Fig. 1. The sensitivity of the system reliability is investigated with regard to the fatigue limit state of prestressing steel. The objective is to identify the most important parameters related to fatigue.

The procedure involves structural analysis, reliability analysis and sensitivity analysis. A simplified approach to analyze the bridge is proposed by Arafah (1986). It is based on the semi-continuum method in which flexural and torsional rigidities of girders and slab are modeled by a system of elastic springs (Jeager and Bakht 1985). The actual response of girders under fatigue loading conditions is calculated, and the cumulative density functions of stress in prestressing strands (stress range spectra) are then developed.

The reliability analysis includes formulation of the fatigue limit state function, statistical description of the parameters and evaluation of the reliability indices for girders. The limit state function is formulated (Al-Zaid 1986, Wang and Yao 1983) using Minors' rule where the safety margin is measured in terms of the fatigue damage accumulation as follows:

$$Z = D_f - D(t) \tag{1}$$

Where

$$D(t) = t\,(ADTT) \sum_{i=1}^{m} \frac{n_i}{N_{fi}} \tag{2}$$

and

$t$ = time;

$D_f$= random variable representing critical damage level at failure;

$D(t)$= random variable representing damage accumulation at time t;

$n_i$= relative frequency for stress range (i) obtained from the developed stress range spectra;

$N_{fi}$= random number representing the number of cycles of stress range (i) which causes fatigue failure of strands;

ADTT= the average daily truck traffic;

m= number of stress ranges obtained from the stress range spectra.

The probability of failure of the $i^{th}$ girder is

$$P_{fi} = P\,[\,D_{fi} < D(t)_i\,] \tag{3}$$

The definition of the system failure involves considerable difficulty. If failure of the bridge is defined as the failure of any of the girders, then the system failure event can be considered as a union of failure events for girders (or collapse modes). There is no general solution for the system probability of failure. However, an exact solution can be evaluated for the case of mutually independent modes of failure (Cornell 1967). The system probability of failure becomes

$$P_f = \sum_{i=1}^{n} P_{fi} \tag{4}$$

where n = number of independent girders (failure modes). For correlated girders (collapse modes) the system probability of failure is dominated by one critical mode. Then

$$P_f = \text{Max } P_{fi} \tag{5}$$

Equations 4 and 5 represent the upper and lower bounds of the system probability of failure. The system reliability index , $\beta_s$, can then be calculated as follows:

$$\beta_s = -\Phi^{-1}(P_f) \quad (6)$$

where $\Phi^{-1}$ is the inverse of the standard normal distribution.

Three errors are considered; incorrect area of prestressing strands, wrong mean of the live load model, and underestimated average daily truck traffic volume. The results show that the system reliability is most sensitive to errors in the area of prestressing strands, and less sensitive to errors in the mean of live load model and the average daily truck traffic. The sensitivity functions, which define the relationship between the magnitude of error and system reliability are plotted in Fig. 2. The calculations are based on a fifty year reference design period.

The sensitivity functions indicate that the fatigue failure depends mainly on the quality of materials rather than loading errors. Therefore, material properties, especially as they related to area of prestressing steel, deserve more attention during design and construction.

## Sensitivity Analysis for a Frame Structure

The second case study concerns a steel frame structure. The frame layout and the applied loads are shown in Fig. 3. Sensitivity analysis was performed to identify the most sensitive parameters. The general form of the ultimate limit state function can be expressed as follows (Moses and Rashedi 1983, Melchers 1983) :

$$Z_i = U_i + E_i$$

$$= \sum_{j=1}^{h} \theta_{ij} M_j - \sum_{k=1}^{p} \delta_{ij} P_k \quad (7)$$

where

$Z_i$ = safety reserve in the $i^{th}$ mode of failure;
$U_i$ = internal random work associated with the $i^{th}$ collapse mode;
$E_i$ = external random work associated with the $i^{th}$ collapse mode;
$M_j$ = random number representing the $j^{th}$ plastic moment capacity;
$\theta_{ij}$ = relative rotation at the j - th location;
h = number of plastic hinge locations;
$P_k$ = random number representing the $k_{th}$ applied load;
p = number of applied loads.

The probability of failure for the i-th collapse mode is

given by:

$$P_{fi} = \Phi \left[ - \frac{\bar{Z}_i(X)}{\sigma_{Zi}} \right] \qquad (8)$$

where $\Phi$ = standard normal cumulative distribution function and $\bar{Z}_i$, $\sigma_{Zi}$ are the mean and standard deviation of $Z_i$, respectively. The upper and lower bounds of the system probability of failure can be evaluated using Eqs. 4 and 5.

Sensitivity to errors in load parameters (P, H) and resistance parameters ($M_p$) are investigated. The developed sensitivity functions are plotted in Fig. 4. The results show that the system reliability is sensitive to errors in applied vertical loads and plastic capacity of frame members. Although the sway mode of failure is sensitive to errors in the horizontal load, it has little impact on the overall system reliability.

## Conclusions

Sensitivity analysis helps to identify the impact that errors in basic parameters have on the structural reliability. The importance of sensitivity analysis from the system reliability point of view is demonstrated by two examples. The developed sensitivity functions provide a basis for the optimization of effort.

## Acknowledgment

The authors acknowledge the support from the National Science Foundation under grant No. CEE-8304962, with Program Director Gifford Albright.

## References

Allen, D.E., "ACI Error Survey Canadian Data," Building Research Note No. 123, National Research Council of Canada, Ottawa, July, 1977.

Al-Zaid, R.Z., "Fatigue Reliability of Prestressed Concrete Girder Gridges," Ph.D. Thesis, University of Michigan, Civil Engineering Department, 1986.

Arafah, A.M., "Integration of Human Error in Structural Reliability Models," Ph.D. Thesis, University of Michigan, Civil Engineering Department, (in progress).

Blockely, D.I., "Analyses of Structural Failures" Proc. Instn Civ. Engrs, Part 1, Feb. 1977, pp. 51-74.

Cornell, C.A., "Bounds on the Reliability of Structural Systems", Jour. of struct. Div., ASCE, Vol. 93, No. ST1, Feb. 1967, pp. 171-200.

Dhillon, B.S., "System Reliability Evaluation Models with Human Errors," IEEE Trans. in Reliability, April 1983.

Frangopol D., "Sensitivity Studies in Reliability-Based Analysis of Redundant Structures", ICOSSAR' 85, Vol. I, Japan, May 1985, pp. 129-138.

Jaeger, L.G., and Bakht, B., "Bridge Analysis by the Semi-continuum Method," Canadian Journal for Civil Eng., 12, pp. 573-582, 1985.

Knoll, F., "Safety, Building Codes and Human Reality," Introductory Report, 11th Conf. IABSE, Vienna, 1979, pp. 247-258.

Lind, N. C., "Management of Gross Errors", Fourth Inter. Conf. on Applications of Statistics and Probability in Soil, Structural Engineering, ICASP-4, Italy, June 1983, pp. 669-681.

Lind, N.C., "Models of Human Error in Structural Reliability," Structural Safety, Vol. 1, 1983, pp. 167-175.

Matousek, M., "Outcomming of A Survey on 800 construction failures," IABSE Colloquium on Inspection and Quality Control, July 1977, Cambridge, England.

Melchers, R.E., "Reliability of Parallel Structural System," ASCE Journal of Structural Engineering, Vol. 109, No. 11, Nov. 1983, pp. 2651-2665.

Melchers, R.E., and Stewart, M., "Data-Based Models for Human Error in Design," ICOSSAR'85, Vol. II, Japan, May 1985, pp. 51-60.

Moses, F., and, Rashedi, R., "The Application of System Reliability to Structural Safety," ICASP-4, Italy, 1983, pp. 573-583.

Nowak, A.S., and Carr, R.I., "Classification of Human Errors," Proceeding of ASCE Symposium on Structural Safety Studies, Denver, May 1985, pp. 1-10.

Nowak, A.S., and Carr, R.I., "Sensitivity Analysis for Structural Errors," The Journal of Structural Engineering, ASCE, Vol. 111, No. 8, August 1985, pp. 1734-1746.

Rackwitz, R., "Planning for Quality - Concepts and Numerical Tools," IABSE Workshop, Rigi Report, June 1983, pp. 49-64.

Ravindra, M., "Treatment of Gross Errors in Nuclear Plant Risk Studies - A Suggested Approach", ICOSSAR'85, Japan,

May 1985, Vol II, pp. 21-29.

Swain, A.D., and Guttmann, H.E., "Handbook of Human Reliability Analysis with Emphasis on Nuclear Power Plant Applications", NUREG/CR-1278, U. S. Nuclear Regulatory Commission, Washington, D.C., Aug. 1983.

Wang, D.F., and Yao, J.T.P. "Fatigue Damage and Structural Reliability," ASCE Annual Convention, Philadelphia, Pennsylvania, May 16-19, 1983.

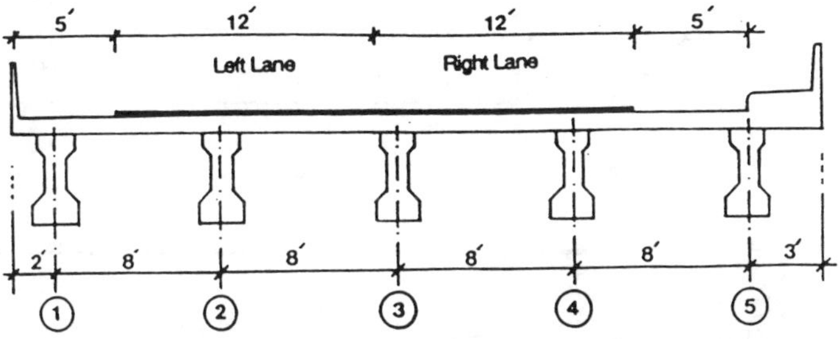


Fig. 1. Mid-span Section of Two Lane Bridge.

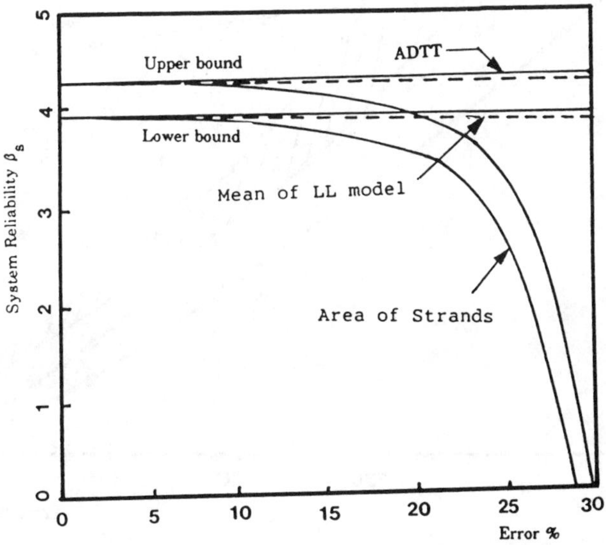


Fig. 2. Sensitivity of Errors in Load and Resistance Parameters of The Bridge Deck System.

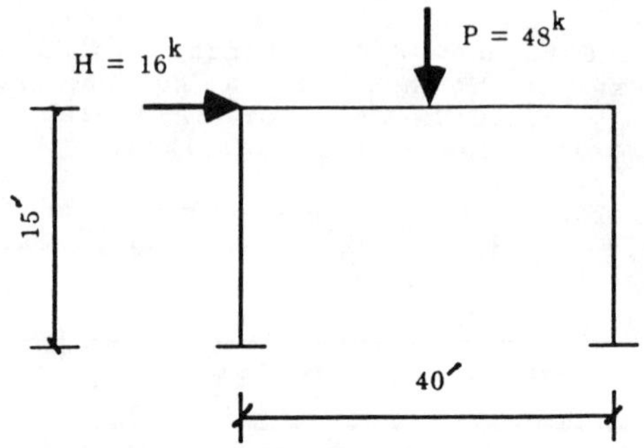


Fig. 3. A Steel Frame Structure.

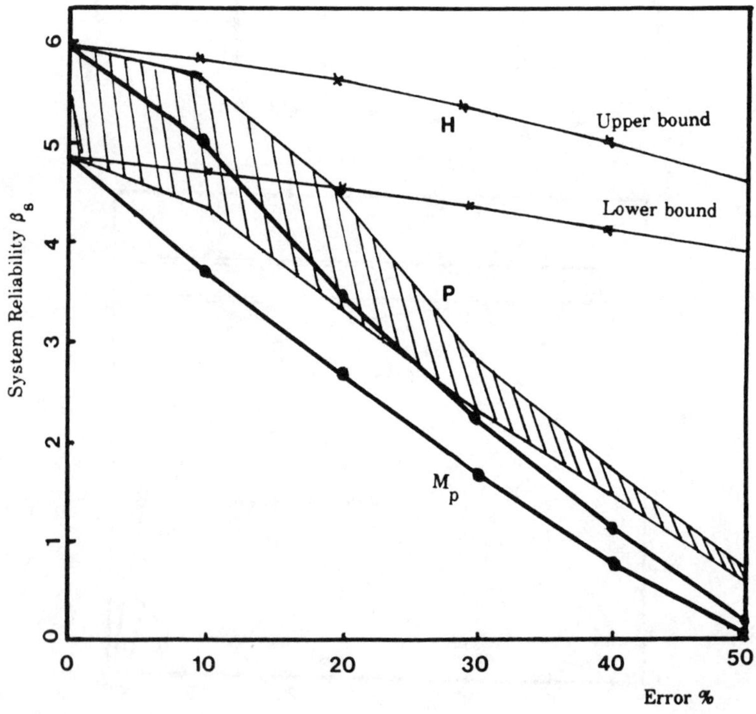


Fig. 4. Sensitivity of Errors in Load and Resistance Parameters of The Frame Structure.

# EFFECTS OF GROSS ERROR ON STRUCTURAL RELIABILITY

W. Brent Hall*, A.M. ASCE

Abstract

The effects of gross errors on structural reliability analysis are investigated through the case history of a gross error problem in chain systems. Some inadequacies of conventional reliability analysis are illustrated and discussed in the context of structural errors. Successful performance of a structure in service is shown to have a proof load effect on the strength estimate of the structure that results in improving reliability, and decreasing probability of gross error, with increasing survival age. The role of quality control in limiting structural errors is examined.

## Introduction: An Example Of Gross Error

In structural reliability a large non-redundant system can sometimes be modeled as a "chain" or "weakest link" system, in which the strength of the system is determined by the strength of the weakest component. A real structure is seldom as simple as a chain, but it happens that the writer was once involved in the reliability analysis of a real chain (Adams 1979), in which gross error played a role. A typical chain was constructed as illustrated in Figure 1, with parallel groups of links (acting as ductile systems) joined in series by pins.

For independent and identically distributed link strengths and pin strengths, it is a fairly simple task to find the distribution of chain strength, if link and pin distributions are known. For example, the equation

$$F_R(r) = 1 - [1 - F_D(r)]^n \, [1 - F_P(r)]^n \qquad (1)$$

describes the cummulative probability distribution of chain strength (R) for n ductile link subsystems (D) and n pins (P). The strength of a group of links is obtained easily as the sum of link strengths. Equation 1 applies to the case of equal-numbered link groups joined by pins, but can readily be extended to the case of unequal link groups life

*Assistant Professor of General Engineering, University of Illinois at Urbana-Champaign, Urbana, Illinois 61801.

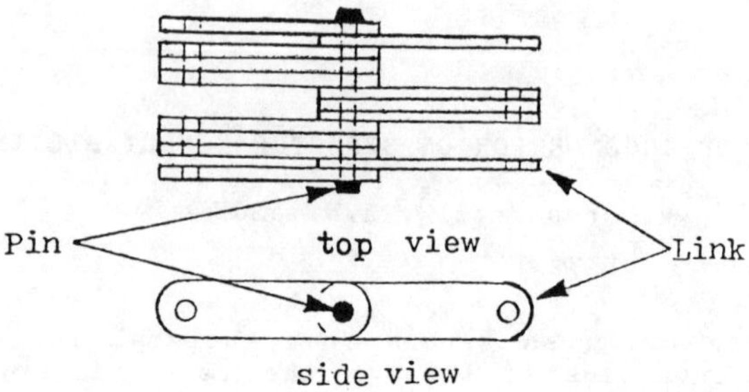


Fig. 1 Link and Pin Elements of a Typical Chain System

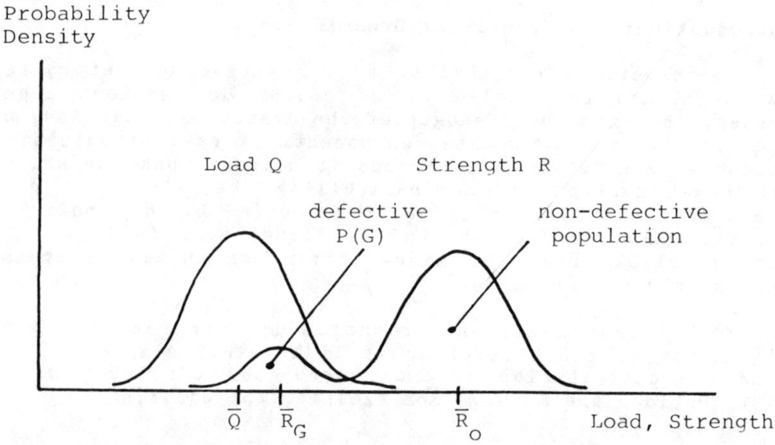


Fig. 2 Load vs. Strength with Gross Error

the chain system shown in the figure.

Thus, by plotting this distribution of chain strength on normal probability paper, for example, it would not be a difficult task to find chain reliability for various loadings using design point calculations.

However, no such calculation proved to be necessary, nor would it have been helpful. Significant numbers of chains were found to be failing in early service at loads too far below the nominal strength to be explained by ordinary statistical variation. The situation was something like that shown in Figure 2, in which a fraction of the population with gross error, P(G), experienced reduced strengths $R_G$, several standard deviations below the nominal (unreduced) strength $R_o$. The probability of failure in this case is

$$P(F) = P(F|G)\ P(G) + P(F|\bar{G})\ P(\bar{G}), \qquad (2)$$

in which G and $\bar{G}$ represent the events gross error and no gross error, respectively. If the statistical information is available, conventional reliability calculations can be used to estimate the conditional probabilities in (Barnett and Hermann 1965). For the chains studied, however, the effects of the gross error were such that defective chains were virtually guaranteed to fail; that is, $P(F|G) \simeq 1.0$. Also, non-defective chains had a very small probability of failure, so that the second product in (Barnett and Hermann 1965) became negligible in comparison with the first. Thus,

$$P(F) \simeq P(G) \qquad (3)$$

that is, for all practical purposes the probability of failure of a chain was simply the probability of gross error.

An investigation of defective chains found that the cause of failure was the occurrence of one or more short pins during chain manufacture, in which case the short pins could not be adequately crimped. Thus, the "connection" between pin and links turned out to be the weak point in the system. The presence of short pins was the result of two types of error. The first, a human error, was the mixing of pins for one type of chain design with the similar pins of another type of design. The second type of error was a machine error that occurred when the wire from which pins were cut bounced back from a stop, causing a pin to be cut short. The human error was dealt with by improved labeling and storage procedures, and the machine error was reduced by a careful adjustment of the machine. However, these measures did not eliminate all short pins, and even with a small probability of a defective pin the probability of a defective chain grew combinationally with the number of pins and links. A good chain occurred only when no pin was defective; thus

$$P(F) \simeq 1 - [1 - P(G)]^n \tag{4}$$

for a chain with n pins each having a constant gross error probability P(G). Thus, one defective pin in 10,000, although a great improvement over the original one defective in 200, still produced 100-link chains with about 1% probability of failure.

Further efforts were therefore taken to reduce the effects of short pins. The final solution involved the invention of a screening device that tested each pin length optically, just prior to insertion of the pin in the chain, and automatically rejected any pin that was too short (or too long). This method practically eliminated the problem of gross error.

Discussion

Because the foregoing example occurred in a manufacturing environment, it was possible to determine that a gross error was present and to zero in on its cause. The designers and builders of full scale structures cannot rely upon mass production to uncover the existence of a gross error and to aid in its elimination. Thus, some features of this example will not apply to structural design and construction. There are, however, some observations that can be made on the effects of gross error that do apply to larger structures as well as the example:

(1) The failure of the "on-paper" safety factor. In this example, the safety levels used in chain design had no impact on true reliability. Indeed, raising the safety requirements on links and pins would only have resulted in improved safety for exactly those systems that did not need it, namely the non-defective systems. In Figure 2, for example, the higher mode of strength would move even further to the right in response to an increased safety factor, while the more significant lower mode, for defective chains, would not move at all. This situation surely applies to many gross errors in structures. Although some kinds of gross errors can be compensated for by a large safety factor (viz. unacticipated loads and the Brooklyn Bridge) there are clearly many others in which no increase in a safety factor will alleviate the error.

(2) An unexpected failure mode. The history of structural failures is replete with this type of error, usually as the result of pushing the design and construction technologies of the time into new and more demanding applications. A good recent book on structural failures is (Petrosky 1985) in which several examples of this and other types of errors are given. The problem for structural designers is, How to design for what is not anticipated? The use of very large safety factors, as mentioned previously, does not appear to be very promising. Rather, it

may be much more appropriate to carefully track the experience of the profession, so that both successes and failures will serve to prove and, when necessary, change design and construction practices. The use of design codes is one way in which the collective experience of the profession is recorded and passed on. Others are clearly needed, so that a designer will know more easily whether he is stepping on new or old ground and, if the latter, benefit from the experience of others.

(3) The irrelevance of complex modelling. In the chain example considered, the relatively complex model of chain reliability was of no help in solving the gross error problem. Only the simplest of probability calculations were needed, and even these had little to do with the final solution. For structures as well, one might suspect that simple models of gross error may turn out to be more useful than complicated ones, and alternatives to conventional probability models may be needed to make use of experience and judgment.

(4) A non-structural solution. It is an important feature of the example considered that the remedies used for human and machine-based errors were not structural. That is, although one solution approach might have been to change the structural design of the chain, the approach taken and found to be effective was to focus on the process of making a chain. So too with real structures, many human and other gross errors may require a process solution rather than a structural solution.

The design and construction of a structure can be viewed as a process in which the performance of human operators (designers, contractors and workers) may determine whether or not the process is in statistical control. A system that is out of control is one in which large deviations from target quality levels (and therefore gross errors) can occur (Deming 1982, Grant and Leavenworth 1980). A production system that is in control is one in which its intrinsic variability cannot be improved without a change to the system. An example of a system change in construction was encountered by the writer when he was involved in the improvement of a weld quality monitor for the U. S. Army Construction Engineering Research Laboratory. The monitor consisted of a microprocessor controller and sensors that measured voltage, current and travel speed of the weld electrode relative to the work. Based on past experience with weld problems, these three measurements allowed control limits to be calculated that, when exceeded, gave warning to the welding machine operator that his process might be going out of control. Process improvements such as this, and other applications of quality control, offer some promising solutions to error problems in structures, although difficulties exist in adapting manufacturing-oriented techniques, such as sample testing, to the construction process.

## Proof Load Effects On Gross Error

One way to control gross errors in structural systems is by proof load testing, a screening test on strength similar to the screening test on pin length used in the previous example (e.g., (Barnett and Hermann 1965, Fujino and Lind 1977, Rackwitz and Schrupp 1985, Shinozuka 1969). One or more structures are tested at a given proof load, and those passing the test are thereby shown to have strengths (in the test mode) greater than the proof load. Assuming no damage from testing, the revised probability density of strength, $f_R''(r)$, is obtained from a truncation of the distribution prior to the test, $f_R'(r)$, at the proof load, $q^*$:

$$f_R''(r) = \frac{f'_R(r)}{1-F'_R(q)} \qquad r>q^* \tag{5a}$$

$$f_R''(r) = 0, \qquad r<q^*. \tag{5b}$$

Thus, if the prior estimate of strength includes a reduction in strength from gross error, the proof load level, $q^*$, can be chosen to eliminate the lower tail of strength uncertainty, including the error effect (Figure 3). Accordingly, revised estimates of strength and reliability are improved after the test. A simple example of proof testing in construction work is the practice of tightening bolts to proof loads during installation. Defective or improperly selected bolts can be eliminated in this way, provided that the installer is aware of the proof strength to be expected from the correct bolts.

An advantage of proof testing is that it does a good job of screening out the worst structures, i.e., those with the lowest strengths. Of course, a proof load test cannot reduce error uncertainties in modes that are not at least correlated to the test mode. Nevertheless, proof testing should not be overlooked as a possible option in human error problems. The proof load test screens out low strength members of the population and increases the reliability estimate of a tested structure.

## Service Proven Structures

Successful past resistance of a strucure to a service load, $Q^*$, acts as a proof test on its strength. In some applications it may be possible that a successful service load and its variance can be estimated from data on existing or past loads, inspections, or other information.

If the value of the proof load has uncertainty, that is, the proof load is a random variable, $Q^*$, then the revised distribution of strength becomes

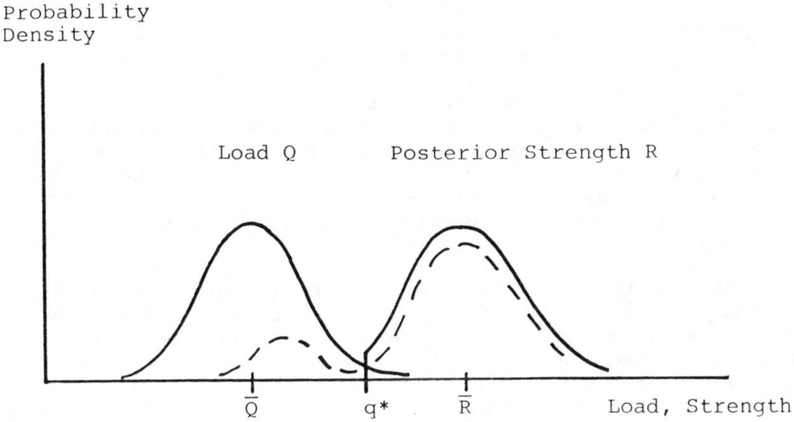


Fig. 3 Screening of Low Strength Structures by Proof Load Testing

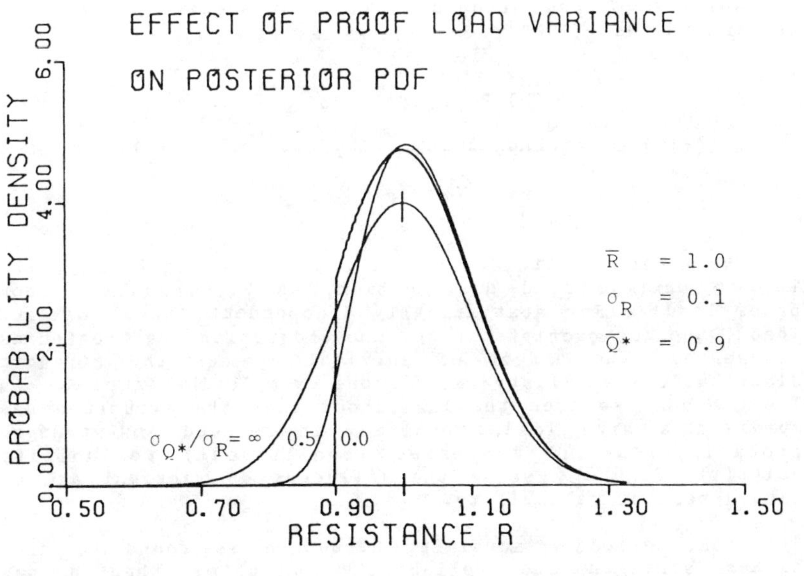


Fig. 4 Strength Distribution after Successful Loading

$$f''(r) = \frac{F_{Q*}(r)f'_R(r)}{\int_{-\infty}^{\infty} F_{Q*}(r)f'_R(r)dr} \qquad (6)$$

This result reduces to Equation 5 when the proof load has zero variance, i.e., when it is known. Equation 6 can be used to revise the initial strength estimate without a formal proof load test (Hall and Nowak 1984). The improved strength estimate is then used to evaluate the structural reliability of the service-proven system for the anticipated design load.

The effect of variance in the proof load is illustrated in Figure 4, assuming a normal prior strength distribution and omitting the effect of gross error. For low variance in the proof load, the probability of low strength components is much reduced, similar to the screening action of a classical proof load test. If the proof load has high variance then little change in the strength distribution occurs.

## Survival Age of a Structure

If a structure has unknown but maintained strength (i.e., R is a time invariant random variable) and survives to an age T under the random load Q, the conditional reliability of the structure for a subsequent t years of service is (Hall 1984),

$$Rel(t|T) = \frac{\int_{-\infty}^{\infty} F_Q^t(r)\, F_Q^T(r)\, f'(r)\, dr}{\int_{-\infty}^{\infty} F_Q^T(r)\, f'_R(r)\, dr} \qquad (7)$$

The above is obtained from Equation 6 using $F_Q^T(q)$, the T year maximum load distribution, as $F_{Q*}(q)$ (the random proof load). For statistically independent annual maximum loads, the superscripts in the above equation are treated as exponents. The effect of survival age on the strength distribution is illustrated for one example in Figure 5. As T increases, so does the likelihood that the structure has survived a high load, resulting in reduced low-strength probabilities and increased (conditional) reliability Rel(t|T). Figure 6 shows the influence of survival age on time-dependent reliability.

Thus, an older surviving structure is found to have higher strength and reliability estimates than a new structure from the same population. The structure experiences a decreasing failure rate throughout its useful life, until the effects of wearout become significant. For this time, an old structure is better than new.

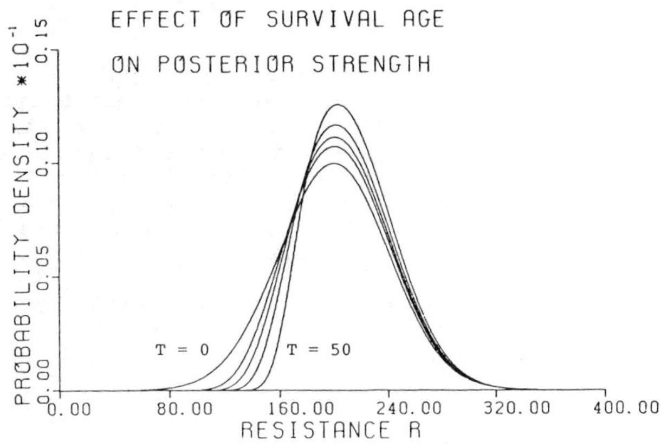


Fig. 5 Proof Load Effect of Survival Age T on Posterior Strength. $\overline{R} = 200$, $V_R = 200$, $\overline{Q} = 100$, $V_Q = 0.3$

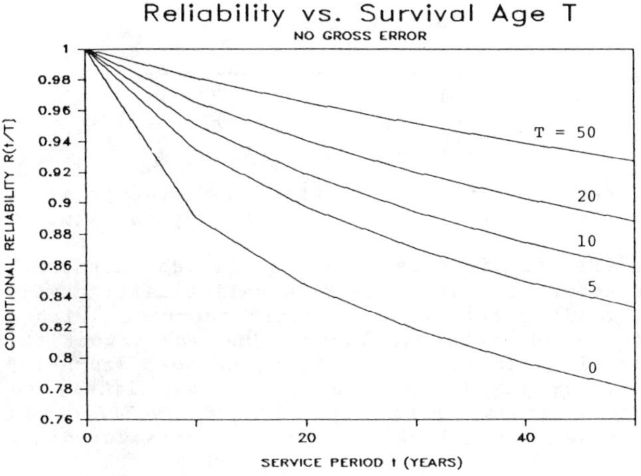


Fig. 6 Improvement in Time Dependent Reliability from the Proof Load Effect of Survival Age T (see Fig. 5).

## Gross Error

A simple model of gross error in strength can be used to show that the effects of gross error on reliability diminish with survival age. Let the population of structures be divided into two parts, one having reduced strength, gR, with probability p(g) and the other having full strength R with probability 1-p(g). g is a factor between zero and one representing the effect of a gross error on strength. p(g) is the probability of gross error.

If R is deterministic at the value r, then the reliability of a structure from this population, conditioned on survival for T years, is

$$\mathrm{Rel}(t|T) = \frac{p(g)\, F_Q^{T+t}(gr) + [1-p(g)]\, F_Q^{T+t}(r)}{p(g)\, F_Q^{T}(gr) + [1-p(g)]\, F_Q^{T}(r)} \qquad (8)$$

Figure 6 illustrates the effect of gross error on time dependent reliability. Possibly the more interesting result is given in Figure 7, which shows the likelihood of a significant gross error diminishing rapidly with service life:

$$p(g|T) = \frac{p(g)\, [F_Q(gr)]^{T}}{p(g)\, F_Q^{T}(gr) + [1 - p(g)]\, F_Q^{T}(r)} \qquad (9)$$

If the unreduced strength is non-deterministic the results are expected to be qualitatively the same.

The effects of gross error on time dependent reliability are characteristic of early life reliability; namely, a high initial failure rate which rapidly decreases with component lifetime, if defects are large. One may expect that, after a reasonable time in service (similar to a "burn-in" time in electronics), surviving structures are unlikely to contain significant gross errors in strength. Effects of gross errors on wearout reliability are not considered here.

## Discussion

The concept of a proof load effect from successful past performance of a structure can be extended further. The success of components or systems in one failure mode affects survival and failure probabilities in correlated failure modes (Hall 1985), and the observed performance of only a

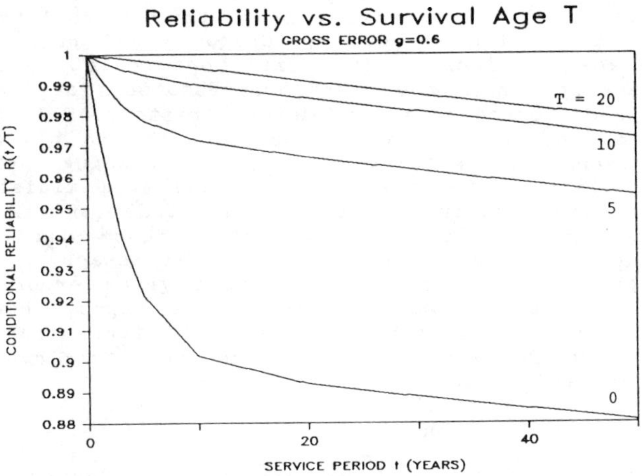


Fig. 7 Effect of Gross Error on Time Dependent Reliability. $g = 0.6$, $P(g) = 0.10$, $R = 200$, $\overline{Q} = 100$, $V_Q = 0.3$

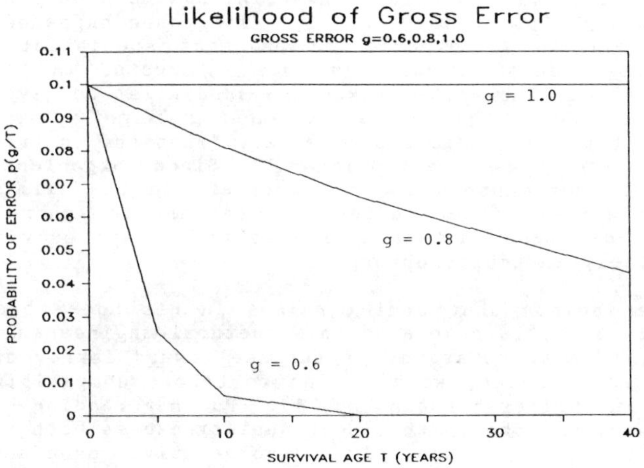


Fig. 8 Effect of Survival Age on the Probability of Gross Error for a Successful Structure. (Same conditions as Fig. 7).

few structures in a series may prove the reliability of a family of correlated structures (Grigoriu and Hall 1984). If one takes the idea to its limit, one could argue that entire sets of design codes and construction practices are in fact proof-tested by the structural history. This brings one back to the importance of experience and to the importance of learning from failures. In the preceding analysis, the probability of gross error was found to diminish with the age of the structure. Although not true for all types of errors this result suggests that to understand gross error, one should study failures, not successes. One is also tempted to conclude that to avoid gross errors, one should copy what has been successful in the past. However, it is not always possible, nor desirable from a creative point of view, for structural engineers to simply replicate past designs and construction methods.

**Management Considerations**

In his recent book on quality control Deming (1982) describes his role in helping to increase industrial productivity and quality in Japan in the 1950's and later in this country. Some of the material presented in the book appears to be relevant to structural design and construction as well as manufacturing. In particular, Deming outlines "fourteen points for management" which are worthwhile examining in the context of gross errors. Some of these points have obvious application to construction. For example, "End the practice of awarding business on the basis of price tag" (point no. 4). The writer once happened to be seated next to an experienced pavement specialist on an airplane. Because a colleague was interested in pavement deterioration, the writer asked the specialist to give, from his experience, the earliest warning sign of serious pavement problems. His response was, "when the contract has been let to the lowest bidder." Since experience and quality of workmanship are so important in the elimination of gross errors, it may often be necessary to ignore price in favor of other factors. This is not always easy to do, particularly in public enterprises.

Other points that Deming raises in his book appear to have only a little revelance to structural engineering. For example, "Create constancy of purpose toward improvement of product and service, with a plan to become competitive and to stay in business" (point no. 1). Further reading on this point reveals a thesis that poor quality causes both re-work and lost sales. Thus, in a competitive manufacturing environment management is motivated by self-interest to improve quality. In the construction environment, however, it is not clear that the same reasoning applies to the elimination of gross errors. Indeed, one even has trouble identifying "management" in the design and construction context. If one regards the structural engineering profession as the managers of risk, then it appears that mechanisms must be found to create the motivation needed to eliminate human errors in design and construction, in the public-interest rather than self-interest environment.

A study of the remaining twelve points made by Deming brings mixed results, along the lines of the two discussed above. Evidently, although the control of human errors in design and construction has many parallels to quality control in manufacturing, there are often special considerations in structural engineering that will require adaptions of quality control methods, or entirely new methods to be developed.

## Conclusion

The effect of gross error on reliability has been illustrated for the case of a chain structure in which a large number elements results in many opportunities for error, namely the occurrence of one or more short pins in the chain. Thus, the reliability of the chain becomes dominated by probability of gross error. Solution of this reliability problem was found not by a structural change, nor by a sophisticated reliability model, but by changes in human- and machine-related aspects of the production process. Neither could the problem have been solved by the use of a large safety factor. This illustration is perhaps not unlike the problem of gross errors in structures, the failures of which also seem to be dominated by the occurrence of gross errors. Elimination of gross errors in structural systems will probably depend on some form of improved process control in design and construction.

The effect of proven reliability on the likelihood of gross error has also been illustrated. The successful performance of a structure in service acts as a proof test on its strength, resulting in a decreasing failure rate and improving conditional reliability with survival age. The probability of a gross error in strength is found to diminish with age, as the structure is verified in service. The results suggest that failures and not successes should be studied for the effects of gross errors, and that the history of structural successes serves to prove out design codes and construction practices.

Some aspects of quality control have been briefly discussed as they pertain to gross errors in structural engineering. Fundamental differences may exist between systems of responsibility and reward in the manufacturing and construction environments. Accordingly, although several techniques of quality control appear to be applicable to structural error problems, management difficulties may arise in the implementation of these techniques.

References

Adams, D., et al, "FMC Pin Sorting Machine", Dept. of General Engineering, University of Illinois at Urbana-Champaign, 1979.

Barnett, R. L., and Hermann, P. C., "Proof Testing in Design with Brittle Materials," Journal of Spacecraft and Rockets, Vol. 2, 1965.

Barry, C., et al, "Weld Quality Monitor", Dept. of General Engineering, University of Illinois at Urbana-Champaign, 1982.

Deming, W. E., Quality, Productivity and Competitive Position, M.I.T. Center for Advanced Engineering Study, 1982.

Fujino, Y., and Lind, N. C., "Proof-Load Factors and Reliability," Journal of the Structural Division, ASCE, Apr., 1977.

Grant, E. L., and Leavenworth, R. S., Statistical Quality Control, 5th ed., McGraw-Hill, 1980.

Grigoriu, M. and Hall, W. B., "Probabilistic Models for Proof Load Testing," Journal of Structural Engineering, ASCE, Vol. 110, No. 2, February, 1984.

Hall, W. B., "Proof Load Effects on Time Dependent Reliability," Fifth Engineering Mechanics Specialty Conference, ASCE, Laramie, Wyoming, August 1-3, 1984.

Hall, W. B., and Nowak, A. S., "Reliability Verification Using Service Loads," Seventh International Specialty Conference on Cold-Formed Steel Structures, St. Louis, Missouri, November 13-14, 1984.

Hall, W. B., "Proof Load Effects on Reliability of Existing Structures", Structures Congress, ASCE, Chicago, Illinois, September 16-18, 1985.

Petrosky, H., To Engineer is Human, St. Martin's Press, 1985.

Rackwitz, R., and Schrupp, K., "Quality Control, Proof Testing and Structural Reliability," Structural Safety, No. 2, 1985.

Shinozuka, M., "Structural Safety and Optimum Proof Load," Symposium on Concepts of Safety of Structures and Methods of Design, International Association For Bridge and Structural Engineering, London, 1969.

# H.E. AND SOCIO-TECHNICAL SYSTEM FAILURE

N.F. Pidgeon*+ and B.A. Turner+

Abstract

The characteristics of engineering as a socio-technical are outlined, and the incubation of failures described. It is argued that the term 'human error' may be properly descriptive of only a restricted class of the human and organisational factors typically associated with failures in structural engineering. Three levels of factor are described: individual, small-group, and institutional.

## Introduction

The growth of engineering has been marked by the influence, and successful application of the physical sciences to engineering problems. This, however, has not been matched by a similar application of the social sciences to the social aspects of engineering problems. Recently, a number of researchers have reported that the causes of many structural failures have been primarily human and organisational (e.g. Blockley 1977, 1980; Matousek 1977; Walker 1981). This awareness of the importance of human factors amongst a small sector of the international structural engineering research community is paralleled by recent developments in other technological domains. For example, following the serious accident at the Three Mile Island nuclear power plant in 1979 a number of human reliability studies have been sponsored (Hanauer 1982), and similar studies might be expected to follow the recent Chernobyl incident.

Neglect of the human and organisational factors influencing technological hazards can result in seriously incomplete assessments of risk (Fischhoff, Lichtenstein, Slovic, Derby and Keeney 1981), and may lead ultimately to inadequate management of safety (Kharbanda and Stallworthy 1983). But before we can move on to build adequate safety assessment models an understanding of the common human and organisational characteristics that lead to failures in the construction industry is necessary.

---

* Department of Civil Engineering, University of Bristol, Queen's Building, University Walk Bristol BS8 1TR, U.K.
+ Department of Sociology, University of Exeter, Amory Building, Rennes Drive, Exeter EX4 4RJ, U.K.

Engineering As A Socio-Technical System

The idea that we are masters of our fate, that we can gather information and act upon it to ensure a surprise-free future, is distinctively a product of Western culture, particularly in its post-Renaissance, post-Enlightenment phase. This cultural view of knowledge as a bridge to possible futures, which we may see and act upon, is at the heart of science"s vision of itself. When our knowledge is utilized to transform the great forces of nature to serve human ends, we have engineering.

Of course, such engineering transformations do not take place in a social vacuum. Individuals, their organisations and groups, and ultimately their cultures, are all implicated in the design, construction, monitoring, operation and maintenance of technological systems. This means that it can be unduly restrictive to talk of large-scale engineering failures (disasters) in purely technical terms. To understand many important issues we would do better to consider the characteristics of a socio-technical system. Figure 1 illustrates the notion of a socio-technical system.

The kernel of the engineering failure problem, and for that matter of successful engineering, is to be found in the interaction of the three basic elements referred to in Figure 1: energy, information, and people. As a result of an extensive analysis of a wide range of contemporary technological disasters, Turner (1978, p. 189) has suggested the following: "Disaster equals energy plus misinformation".

This observation points out that inadvertent or unintended transformations of the world arise when energy is misdirected as a result of inadequate information. A corollary of this is that the avoidance of disaster can be achieved by replacing misinformation with perfect information. To reach this point we need only to aspire to a condition of omniscience! We can, of course, obtain full information, omniscience of sorts, in some closed systems. However, we must recognise that for most practical purposes we operate in a contingent universe in which there are ultimate limits upon our ability to reduce uncertainty, or to master all of the open-ended and perverse qualities of our environment. If our understanding of physical systems is imperfect and uncertain, then this is doubly the case when we consider human and organisational factors.

If we learn to think of engineering systems in socio-technical terms we become critical of approaches to the causes of disasters which stress only the technical factors. Such approaches result in restricted learning about hazards, a point that we probably do not need to reinforce to the participants at this workshop. A more thorough learning experience, and further reduction in the incidence of failures, can only arise from a consideration of the technical, human and organisational causes of disasters. Such a programme might profit, as in the case of our own current project (Blockley, 1986), from collaboration between engineers and social scientists.

It is commonplace to ascribe the cause of an accident to "human error", but if we do so we do not adequately recognise the complex character of socio-technical systems. That is, that while failure can often be ascribed to human and organisational influence, the term "human error" ought to be deployed only with considerable caution when describing such factors. The notion of "human error" is important in Western societies which have a longstanding habit of allocating blame on an individual basis. But the attempt to find an identifiable culprit may well, as we shall see later, be at odds with the more subtle causes of a failure, which are often complex, multiple and rooted in properties of the socio-technical system.

For example, an accident might be ascribed to a simple piece of misinformation, such as an error in a design calculation. However, a diligent accident investigator, taking a socio-technical perspective, might ask a range of questions to place this error in context. Why did the error occur? Was it due to a single individual or a team? Where did the information needed to correct it reside, and was it available to the designer? Was the designer adequately trained to deal with this problem? Did the designer have the time, resources and staff needed? Was the error a personal slip, or was it the product of a standard procedure being utilised in a non-standard situation (and should the designer have recognised this)? Did procedures exist for checking calculations? Who might have ensured that these checks did take place? The answers to these and other questions are all relevant if we wish to discover why a socio-technical system has malfunctioned, and such answers force us not to look at "human error" as a simple undifferentiated phenomenon, but to see it as a family of problems of behavior, which might have their origins in managerial activities, in professional information systems, in communication between professionals and contractors, in personal misunderstandings and misapprehensions, as well as in individual failings or lapses. When we recognise the complexity and variety of the sources of our difficulties, we are better placed to respond to them as designers, and as managers or educators in design and construction.

## The Incubation Of Disasters

Analysis of case-studies of many different types of disasters shows that we are not justified to blame such events upon "Acts of God". Many large-scale disasters display similar characteristics and patterns. It is difficult to predict these events precisely, but after the event the patterns of build-up recur again and again (Turner 1978).

Consideration of disasters suggests that three stages typically precede the onset of any significant failure, and these are illustrated in Table 1. No more than an outline of this model can be given here. In Stage I the situation is nominally normal. Those responsible for the technological system operate with an adequate set of beliefs about accepted hazards, and appropriate precautionary norms such as laws, codes of practice, and "engineering judgements". Stage II is defined as the hazard incubation period. Typically this is associated with a set of events, compounded over time, that together produce an unnoticed situation at odds with the accepted beliefs and norms of Stage I. Finally, in Stage III the irrecoverable onset of the disaster follows a final precipitating, or trigger event. The trigger might be a final critical error, or a slightly abnormal (but within design limits) operating condition. An example of a disaster development sequence, leading to a multiple failure of roofs under snow loading, is reported in Pidgeon, Blockley and Turner (1986).

The notion that separate undesirable events compound in an incubation period is one that is not entirely new to structural engineering. For example, Pugsley (1973) introduces the terms "proneness" to failure, and the "climate" in which an accident is set. This suggestion has been extensively developed by Blockley (1977, 1980) in his structural safety studies. Knoll in a biological metaphor of the building process, notes that "human error , like any parasite worth its name, will attack in particular the portions of the organism that are already weak or sick and there it will thrive" (1980, p. 1060). And more recently, Lind has observed that failure is the result of multiple "human errors": while singular errors rarely cause problems "it would take the coincidence of several errors of design, construction or use to precipitate failure" (1983, p. 174). One outcome of this is the long, complex, and sometimes indecisive, litigation that may follow any failure, as the various parties involved in a complex system attempt to concentrate blame at one point or other in the system.

In order to understand such complex systems, we need to study the social context within which engineering occurs. Analysis of this requires an understanding of both individual, psychological, and organisational and cultural, sociological factors. When we look in this way at the human patterns of professional engineers, we are better placed to realise that safety in one field of engineering, such as the construction industry, might be profitably based upon experience from a range of other engineering specialisms, because many of the human and organisational sub-components of socio-technical systems (Figure 1) are common to all engineering disciplines. What principally distinguishes the different branches of engineering (e.g. structural versus mechanical) is the form, and scale, of the energy and artifacts manipulated, rather than the systems of professional activity. Hence, while the consequences of failure vary between the different branches of engineering, it is possible to learn from the fallibility of others, as well as our own.

Two other significant implications arise from the three stage model. First, the successful detection of a potential incubating hazard before a trigger event occurs brings to our attention a near-miss, and in principle we can learn as much from a near-miss as from a complete failure. In some settings of public concern this has led to the introduction of near-miss feedback reporting mechanisms; for example, as in the civil aviation industry. Second, whilst the precise prediction of the where and how of any specific failure might be impossible (the particular combination of events, and uncertainties being unique in their manifestation, if not their general characteristics), this does not mean that we cannot prevent some failures. Specifically, prevention may be attained if we can identify an incubating hazard at an early stage. In some settings, to use Pugsley"s (1973) phrase, an index of "proneness to failure" might be required.

## Factors Underlying Failure

The task of identifying those patterns of events which signify an incubating disaster presents a number of significant difficulties. For example, how is the effect of hindsight to be overcome, and how is a true incubating hazard to be distinguished from the "normal" difficulties that are associated with any typical construction project? These are questions which will not be discussed here, save to note one methodology that is currently under development at the Universities of Bristol and Exeter (Blockley 1986).

The aim of our concluding discussion is more limited: it is, specifically, to outline the common categories of event that are typically associated with incubating hazards. By this means it is hoped to throw some light from the perspective of social science, upon events that are typically labelled "human error".

The notion of an error implies a deviation from some assumed "normal" state of affairs. Such a view presupposes that the norm can indeed be unambiguously defined. Much of the legal process following a structural failure is devoted to "expert witnesses" attempting to define what normal actions would have been; for example, common good practice, or the interpretation of codes of practice. In this way deviation from the norm is established, often as the customary Western definition of professional negligence. However, it is significant that expert witnesses often disagree as to the appropriate norm in many situations. The difficulty of establishing normative criteria where ambiguities exist with respect to intentions, with respect to procedures, and particularly with respect to anticipation of the future, is well known in sociology. These difficulties are compounded when we attempt to establish patterns in hindsight, as is usually the case when we are trying to pin down a "human error".

For current purposes purely technical factors, such as scale-up effects, are disregarded. Neglecting these there remain three distinct categories of event which contribute to the incubation of disasters; individual, small-group and institutional. These are the principal routes for the introduction of misinformation into the building process.

a) Individual Factors

The study of the interaction between individual humans and machines is the domain of ergonomics, latterly titled engineering psychology in a recent review by Wickens and Kramer (1985). Much of ergonomics is predicated upon the assumption that the human individual in a complex socio-technical system can be regarded as no more or less than another single component; one that either performs, or fails to perform, some well-defined function. Insofar as individual functions, or goals, can be well specified, we believe that it is at this level of analysis that the term "human error" can be most usefully utilised. For example, the psychologist Norman (1983), makes a distinction between slips or lapses, and mistakes. A slip or lapse can be defined as failure to achieve a planned goal, such as an arithmetic error in a design calculation. A mistake is made when the plan of action is itself inadequate, and an inappropriate goal is correctly met, as when the arithmetic is correct but the calculation contains an inappropriate conceptual assumption. It is at this level of activity that human reliability studies (e.g. Melchers and Stewart, 1985; Swain and Guttman, 1980) have had modest success in attaching probabilistic estimates to error rates, although as Reason (1985) has correctly pointed out, the human reliability approach cannot as yet deal with the important cognitive activities of decision-making and problem solving.

b) Small-Group Factors

The individuals involved in socio-technical systems do not proceed with their tasks without social controls. All will be members of a least one, and probably several, small-groups with either a direct or indirect input to current projects. For example, a designer will be a member of the small-group of colleagues in the design office, and probably also of a group that meets on site to discuss project progress. At a greater distance the designer may be a member of a local professional body. Although it is, at first sight, an easy question to ask what the goals of a small-group ought to be, this is often less than clear, particularly when the group is part of a much larger organisation (e.g. March and Olsen 1976). We might ask whether it is the aim of a design team to meet a client"s requirements in full, to make profits for the design practice, to build a safe and economic structure, to educate the junior members "on the job", or a combination of all of these? Such requirements should not necessarily be expected to be in complete harmony. If the functions of a group are ambiguous, then the question of when a "collective error" has occurred is equally so.

High on the list of small-group problems are those of ensuring adequate communications and work role definitions. Communication problems are a major source of misinformation and potential mishap. For example, one of the contractor"s employees, for a variety of reasons, may be working from an outdated drawing. Unless this is checked, and rectified, the design engineer"s vision of the final structure will not accord with what is built, with possibly serious consequences. Furthermore, a lack of adequate definition of roles, particularly in relation to responsibility for safety during the erection of a structure, can lead to major difficulties. Precisely how such factors are to be modelled is an open question.

c) Predisposing Institutional Factors

This third level of factors relates to the operation of the construction industry as a whole, together with its associated institutional bodies and organisations. The small-groups of the construction industry do not associate randomly; they are tied together by rules, managerial edicts and by custom and practice. Taken together, they constitute a slowly changing cluster of social arrangements, which has its own values, beliefs, outlooks and practices. At this level, beliefs in the industry about hazards, or about the need or otherwise for changes in safety-related practices, develop and modify in ways which are influenced by individual professionals and small work groups. Such beliefs also, in turn, constrain the work activities of those in the industry. Normative rules which may be specified as codes of practice, or which may remain more tacit, provide some image of "appropriate professional behavior" for the individual engineer. Unfortunately, as with small group goals, the desirable aspects of professional practice or inter-professional relations are rarely expressed precisely, and such clarity as exists may often be at the expense of a lack of clarity about other obligations. For reasons such as these, it is often more appropriate to talk of system failures rather than errors when we are looking for the origins of a particular inadequacy at this level.

One major problem with any set of norms, such as the

rules embodied in a code of practice, is that, however rigorously they are formulated, they are unlikely to cover all possible cases. The individual design engineer may recognise that a situation is indeed significantly atypical, and that it cannot be dealt with on the basis of the code-specified criteria. In the United Kingdom the design engineer"s freedom in this respect is a provision of the Building Regulations; the engineer is charged with the responsibility of utilising different criteria, such as higher loadings, where he or she deems this to be necessary. However, the successful detection of a significantly atypical case may not be easy. In the case-study already mentioned (Pidgeon, Blockley and Turner 1986) a loading code"s lack of provision for asymmetric snow loading patterns was one of the factors leading to several roof failures in the United Kingdom. Blockley (1980) has illustrated the difficulties of dealing with such uncertainties.

## Concluding Comments

The characteristics of engineering as a socio-technical system have been outlined. It has been noted that large-scale failures, or disasters, are rarely caused by any single factor. Rather, disasters "incubate" over a period of time as a number of causal factors interact. Three levels of causal factors have been discussed: individual, small-group and institutional. This analysis suggests that while disasters are indeed failures of our "social engineering" (cf Stech 1979), the term "human error" may be properly descriptive of only a small subset of the potential non-technical causes. The analysis also suggests the need for innovative methods, to complement and extend those of human reliability analysis, for the assessment of risk, and the better management of safety.

## Acknowledgement

This paper was prepared with the financial support of a grant from the Joint Committee of the Science and Engineering, and Social Stiudies Research Councils of the United Kingdom

## References

Blockley, D.I. (1977), "Analysis of Structural Failures", Proceedings of the Institution of Civil Engineers, 62. 51-74.

Blockley, D.I. (1980), The Nature of Structural Design and Safety, Chichester, Ellis Horwood.

Blockley, D.I. (1986), "Towards an A.I. Management Tool to Aid in the Control of Structural Safety", Paper prepared for National Science Foundation Workshop, Modelling Human Errors in Structural Design and Construction, Ann

Arbor, Michigan, June 1986.

Fischhoff, B., Lichtenstein, S., Slovic, P., Derby, S.L. and Keeney, R., (1981) Acceptable Risk, Cambridge, Cambridge University Press.

Hanauer, S.H. (1981), "The Human Factor in Nuclear Power Plant Safety: Progress Since Three Mile Island", Progress in Nuclear Energy, 10, 299-347.

Kharbanda, O.P. and Stallworthy E.A., (1983), How to Learn from Project Disasters, Aldershot, Hants, Gower.

Knoll, F. (1980), "Pursuing Human Errors", Proceedings of the 11th Congress of the International Association for Bridge and Structural Engineering, 1056-1062.

Lind, N.C. (1983), "Models of Human Error in Structural Reliability", Structural Safety, 1, 167-175.

March, J.C. and Olsen, J.P.(1976), Ambiguity and Choice in Organizations, Tromso, Norway, Universitetsforlaget.

Matousek, M. (1977), "Outcomings of a Survey on 800 Construction Failures", International Association for Bridge and Structural Engineering, Colloquim on Inspection and Quality Control, Cambridge, July 1977.

Melchers, R.E. and Stewart, M.G., (1985), "Data-Based Models for Human Error in Design", ICOSSAR 1985, 4th International Conference on Structural Safety and Reliability, II, 51-60.

Norman, D.A. (1983), "Position Paper on Error", Nato Confer on Human Error, Italy.

Pidgeon, N.F., Blockley, D.I. and Turner, B.A., (1986). "Design Practice and Snow Loading: Lessons from A Roof Collapse", The Structural Engineer, 64A, 67-71.

Pugsley, A. (1973), "The Prediction of Proneness to Structural Accidents", The Structural Engineer, 51, 195-196.

Reason, J. (1985), "Predicting Human Error in High-Risk Technology", Paper presented at British Psychological Society Annual Conference, Swansea, March 1985.

Stech, F.J. (1979), Political and Military Intention Estimation, Bethesda, Maryland, Mathtech Inc.

Swain, A.D. and Guttman, H.A. (1980), Handbook of Human Reliability Analysis with Emphasis on Nuclear Power Plant Application, NUREG/CR - 1278, Washington DC, GPO.

Turner, B.A. (1978), Man-Made Disasters, London, Wykeham.

Walker, A.C. (1981), "Study and Analysis of the First 120 Failure Cases", Structural Failures in Buildings, London, The Institution of Structural Engineers.

Wickens, C.D, and Kramer, A. (1985), "Engineering Psychology", Annual Review of Psychology, 36, 307-348.

Table 1, The Sequence of Events Associated with the Development of a Disaster

| | | |
|---|---|---|
| Stage I | Notionally Normal Starting Points | |
| | (a) | Initial culturally accepted beliefs about the world and its hazards. |
| | (b) | Associated precautionary norms set out in laws, codes of practice, and 'engineering judgements'. |
| Stage II | 'Incubation Period' | |
| | The accumulation of an unnoticed set of events which are at odds with the accepted beliefs about hazards and the norms for their avoidance. | |
| Stage III | Trigger Event | |
| | Final critical event or abnormal operating condition. Irrecoverable onset of disaster and ultimate transformation of general perceptions of Stage II. | |

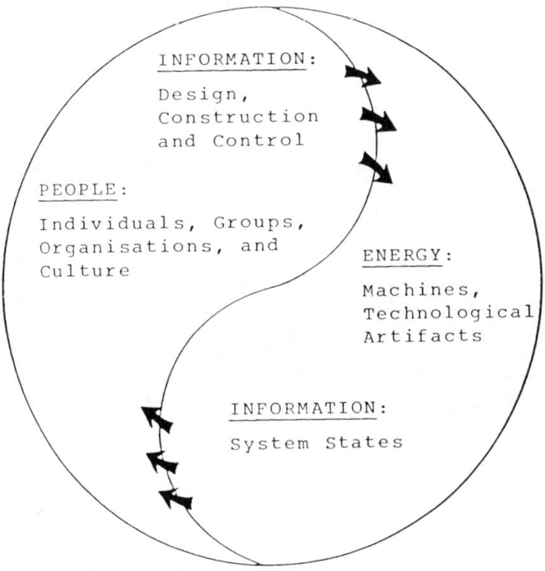


Fig. 1 Engineering as Socio-Technical System.

## REPORT OF THE WORKING GROUP ON OCCURRENCE OF ERRORS

Memebers: Robert E. Melchers, Co-Chairman; Neal FitzSimons, Co-Chairman; Gita Amiri; Colin Brown; Owen G. Ingles; Franz Knoll and Nick F. Pidgeon.

Definitions

Error is defined as a noncompliance with code, or more generally, a significant deviation from standard practice.

Failure is defined as a lack of meeting expected performance level. It can be measured in terms of consequences (cost or utility). Failure can be defined as a significant loss of utility. Survival needs of engineering professionals tend to set the levels of expectation used in design and construction. These may be different from external expectations.

Why are we discussing human error?

There are two conflicting notions:

1. The current average risk of failure seems to be at a level which is acceptable by society. Acceptable risk level depends on knowledge of risk and its perception by a significant number of members of the society. For individuals and projects (or groups of projects), the acceptable risk levels depend on funds available.

2. Legal liability costs have increased, which suggests that the society is not satisfied with the risk levels. This is due to litigation pressure from clients and lawyers. However, litigation activity does not necessarily reflect society wishes or perceptions. We are witnessing a change in the legal profession's perception of its capabilities for its clients and its ability to deliver. The situation is shown in the flowchart below.

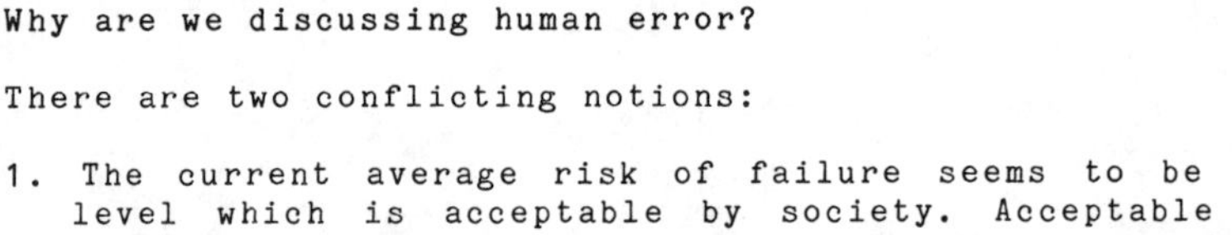

Consequences include loss of stature, loss of business, increased insurance, increased probablity of lawsuits. It is important to study the relationship between competition and quality (not always evident), and between competition and errors (requires feedback which profession does not appear to want).

To reduce consequences, the engineering profession has largely responded by a call for error control which leaves existing legal, educational, etc. boundary conditions untouched. This may not be the `global optimal' perspective or action. Large institutions (eg. US government) are walking away from excessive insurance to cover losses themselves. As a consequence, we are getting opulent structures due to reduced risk level. This is also due to insurance premium problems (eg. 'no claim bonuses`) and costs of litigation.

General Discussion

It is important to consider the environment in which errors occur. An engineer operates as an individual. His motivation is based on self-preservation, survival, competition, distraction, and self advancement. Professional considerations come secondary, and Society comes third.

An important source of error is procedural. This can involve the management structures used in the control of a project and the communication systems set up between different disciplines and parties.

A developing pattern is the increase in the size and complexity of the specification contract documents. This has arisen from fear of responsibility. By delegating responsibility in the contract documents, architects and engineers are avoiding liability. This situation has created several problems. Not all responsibility may be assigned. Some items may be ignored (eg. temporary structures). A complex document creates confusion and ambiguity, which increases error likelihood.

Structural engineers are caught in a squeeze. The liability insurance costs are going up; on the other hand the relative pay and rewards are going down. Engineers are faced with more lawsuits. The pressures are not necessarily caused by an increase in the number of human errors, although a perception of increasing errors exists. This is because the designers have received more legal responsibility and blame. In the past, such a blame may have been placed on an `act of God'.

Another reason engineers receive more blame is that society wishes to reduce risk, without increasing the funding for additional safety. This brings out the central theme of this Workshop, which is to increase the reliability of structures without increasing their costs.

The idea of risk acceptance varies from one time to another and between different societies. For example, projects carried out in the national interest during war-time, or in the developing nations, often involved a higher risk than what is acceptable today. A Challenger spacecraft blew up recently. Naturally, it was taken as a serious failure by the people of the U.S. An argument arose as to whether this accident was important because the media made it so or because the people had an emotional interest in it. This argument cannot be resolved, but in any case perception and perspective play a role.

Two types of errors were considered: opulent errors (overdesign, creating unnecessary cost) and stingy errors (economizing so much that future errors may occur due to premature deterioration).

The group discussed what might be done to reduce errors and the consequences of errors. If reduction of consequences is the goal, changes may be made in public opinion and in the tort system. Engineers could devise systems which are error survivable, ie. redundant structures.

Methods to eliminate errors were discussed and several suggestions were made. A point was made that engineers have become over zealous in modeling extreme natural events. Design events are predicted using computer models and historical data. This gives the public the opinion that engineers have everything under control. Hence, the responsibility is placed on the engineer.

It has become apparent that many problems or errors develop over time. In this instance, the owner plays a major role in error mitigation. It was suggested that structures be provided with owners manuals and the warrantee be contingent upon the owners' actions.

In a related idea, engineers must be trained to think of their designs as total systems, surviving 50 or even 100 years. This training should take place at the undergraduate and professional level. The concept of preventive maintenance falls under this category.

The most important idea put forward is the need to study the parameters involved in the design and construction process. The intention would be to determine those items which made the difference between a successful and unsuccessful project. In order to determine the influencial factors, data would have to be collected from both successful and unsuccessful projects. Apparently, several case studies exist for projects in which accidents have occurred but none for well run projects. These successful projects would serve as the control group in the experiment.

Many other factors can be studied for their correlation to success. Among them are the engineer selection process, project management structure, degree of inspection, and education of engineers. Such studies would provide the basis for a critical review of human error and the development of new models.

Error Ocuurence

Gross figures need verification. Some errors are only gradually discovered. However, in many cases the same errors are repeated again and again.

Error Mechanisms

- Procedural matters are important (eg. design-construction interface, planning of design)

- Buckpassing; ie. transfer of responsibilities/actions.

- Interprofessional communications.

- Organizational links (official vs. actual).

- Record maintenance, eg. `as-built' records.

- Information overload, eg. huge specifications result from attempting to cover all possibilities.

Human Factors and Other Factors

- Broaden data collection to cover Pugsley's factors.

- Investigate `successful' projects.

- Effect of professional registration on error (or quality).

- Effect of inspection (on-site) on error (or quality).

- Effect of fast track construction on error (or quality).

- How to `do' this kind of research?

- Effect of `on-site' engineer (and his role).

- Procedures for safe construction work and structural safety during construction.

- How to design fault-tolerant structures (load sharing, fail-safe).

- Environment of work process - eg. checklists, CPM for critical stages/parts (need data), peer review effectiveness.

- Selection processes (selection of engineers, registration of engineers, self-selection of students).
- Manual for building (project), life time monitoring (more responsibility).

Other Issues

- Monitoring of projects - gradual deterioration, optimal inspection strategies.
- Education - probabilistic thinking, design/construction cases, cooperative programs, convergence of research results on human error.
- How to implement the required research projects.

# REPORT OF THE WORKING GROUP ON STRUCTURAL RELIABILITY MODELS

Members: Ruediger Rackwitz, Co-Chairman; Niels C. Lind, Co-Chairman; Abdulrahim M. Arafah; Dan Frangopol; Mircea D. Grigoriu, Janusz W. Murzewski, Andrzej S. Nowak; Norris Stubbs and Y-K Wen.

## Discussion

### 1. Hazard Scenarios

A scenario is a sequence of events that may lead to failure, a hazardous connection, an event tree with added branches for possible errors. For further discussion see "Scenario Methodology in Natural Hazards Research" by N.J. Ericksen (University of Colorado, Boulder, 1975).

Analysis of hazard scenarios is a device for control and a modeling tool. It helps to avoid future errors.

Event tree approach should be extended to include error generation, detection, and elimination.

The Group was not able to generalize the scenario approach from known failures.

It was recommended that ASCE organize a Task Committee to consider the scenario approach (continuation of the work presented in J. Schneider' report).

### 2. Sensitivity Analysis

Sensitivity analysis is an effective tool to control some of the predictable errors. It may be useful in preventing a progressive collapse.

Sensitivity analysis should be extended to include a proper cost-benefit analysis of allocation options.

The models should include case studies based on practical data (failures, near failures, error surveys).

Sensitivity analysis applied to design codes will help to prioritize the needs for further research and development.

### 3. System Reliability

In reality structures are not ideally elasto-plastic.

Therefore, a high degree of redundancy is needed to prevent a progressive failure, unless the structural components are ductile. Failure of one brittle component may trigger the overall collapse.

In brittle systems, it seems that neither safety margins nor degree of redundancy are effective strategies to counter human error.

Influence of gross errors on progressive collapse models and system reliability is an important almost unexplored area of research.

4. Data Bases

They should be designed to produce the data needed by reliability models.

A purely statistical question: we are so far from the random sampling situation. What does this censored and filtered data really mean without reference to a population?

5. Reliability Analysis and Human Error

Should reliability analysis include human errors? The answer depends on the purpose of the analysis. The profession is collectively responsible for performance of structures, and designers are individually responsible.

6. Other topics

Various methods of reliability analysis were considered from the human error modeling point of view (Bayesian vs. fuzzy methods).

Future Research

- Scenarios.

- Statistics of errors, frequency of errors, estimation of failure rates, consequences of errors. - Sensitivity measures (cost-benefit optimization).

- Event tree concept.

- Progressive collapse problem.

- Checklist.

- Allocation of limited checking time.

- Fuzzy methods in error modeling - do they help?

- Errors during construction.

Recommendations

1. It is recommended that ASCE organize a Task Committee to consider the application of the scenario approach in evaluation of the structural reliability and modeling human errors. This initiative requires a multidisciplinary expertise including structural reliability, statistics, design, construction, forensic engineering, and other areas.

2. It is recommended that the profession, or code writing bodies (AISC), provide public domain software for error free design (i.e. design in conformity with the code).

Both recommendations were unanimously supported by the participants of the Workshop.

## REPORT OE THE WORKING GROUP ON ERROR CONTROL STRATEGIES

Members: David I. Blockley, Co-Chairman; Ted V. Galambos, Co-Chairman; David E. Allen; Robert I. Carr; Fabian C. Hadipriono; W. Brent Hall; Juan A. Melgarejo; M.K. Ravindra; Heki Shibata; and Carl J. Turkstra.

### Discussion

1. In order to control errors, there is a need to develop our understanding of the nature and cause of errors, so they may be categorized and described. This should not be taken to the deepest levels of analysis (at least not by civil engineers), but to a sufficiently deep level to enable control strategies to be formulated.

2. An error is an event which is outside normally accepted practice. In a sense, all error is human error, and all engineering practice is a process. An error could also be defined as the process going out of control.

3. Errors can be classified as those of concept, those of process, and those of intent. Errors may occur in the activities of individuals, small groups, or of institutions (large groups).

   Examples of concept, process, and intent errors for an individual:

   Failure: A washer was missing from a bolt.

   Explanation:

   Concept - The need for a washer was not understood by the designer.

   Process - The instructions to fit the washer were not followed (unintentionally?).

   Intent - The instructions to fit the washer were understood, but were intentionally not carried out.

4. Professional practice is believed not to be in particularly good health. One simple indicator of this is the amount of current litigation. There seems to be a breakdown of trust in professionally responsible decision making. This is an extremly complex problem to which there is no easy solution. The following discussion will

be relevant to the improvement of this situation.

5. Our goal is to reduce the consequence of errors. In order to do that, we will need a deeper understanding of the nature of errors, so that they may be detected and remedial actions can be taken. The extent of the detection and remedial work will depend on the consequences of error.

6. There are some immediate actions which can be taken which would have significant impact on reducing errors.

   a) Teach students to think about thinking (e.g. de Bono's Practical Thinking).

   b) Develop case histories of errors in projects to illustrate the potential and importance of errors. Publish these cases for study by practicing engineers.

   c) Improve the teaching of design by modeling the design process as a decision making tool.

   d) Develop better management training for engineers. Improve the awareness of individuals and groups by using techniques which are already identified by management scientists.

   e) Develop fact sheets or standard action sheets which describe standard items and known problems. Include causes and possible prevention measures, with examples (these may eventually be expert systems).

7. In order to develop long term strategies for basic research, three important factors should be considered.

   a) The need for interdisciplinary research work, involving other disciplines with the problems in civil engineering. The mechanism for funding such work will need to be examined.

   b) Recognizing the problems of defining the methodology of work. How experiments can be set up to test hypotheses and what can be measured.

   c) Consider the attitudes of the organizations in the construction industry and the attitudes of the public. In particular, the attitudes and expectations of the client are crucial. If, by comparison with other engineering industries, construction remains a low budget industry, then the financial climate is bound to strongly affect the possibility of error.

8. The following research areas are thought to be important.

a) A review of the organization of construction in various countries throughout the world. This review would investigate the relationship between the systems of error control and the cultural attitudes and legal systems involved. The review would also include the role of codes and specifications and the procedures for checking.

b) Work with organizational behaviorists to study organizational problems and identify possible hypotheses which would be tested via suitable experiments. Identify what can be measured and in what way.

c) The development of "Appropriate/Adequate" Quality Control/Assurance programs. QA seems to imply many possible levels of activity. For low consequence problems, minimal QA procedures may be defined. For high consequence problems, there are a number of analytical tools which could be developed, for example:

(i) Decision support systems, with particular emphasis on the analysis of sparse data and imperfectly understood problems. We need to relate process alternatives to their consequences in order to determine most effective strategies.

(ii) Hazard scenario analysis.

(iii) A theory of checking which would involve the identification of priorities of elements for checking. The role of proof load testing, the design of inspection programs, fail safe design, the design of observation programs.

(iv) The development of computer programs to ensure standard elements are designed according to the codes. The evaluation of the impact of code changes on other codes (e.g. effect of changes in ASTM standards upon AISC code assumptions).

d) Data collection for case studies of success, failure and near misses. Use these case studies to develop models for error detection. Include in the models the "climatology" of accidents and the communication problems involved. Care must be taken in the methodology of data collection. Problems at the institutional level should also be included.

e) Study the ways in which attitudes are developed within an industry and how organizational systems may be adjusted to improve incentives for individuals and groups.

f) Study the interaction of the construction industry, the legal system, and the insurance industry, with regard to error control. Determine better ways to distribute responsibility to ensure that all items are handled responsibly.

g) Apply management science methods to develop decision models. These decision models can then be applied to the entire system of design and construction.

9. It was proposed and approved by the group, that a fixed percentage of money be taken out of the total contract price of structures and be used for the purpose of design and construction research and development.

Summary:

In order to reduce errors, we must gain a better understanding of their cause and consequence. This will require the classification of errors by type and level of occurrence. Levels may range from individual, to small groups, to institutions. Errors must be studied in the context of the circumstances in which they have occurred: management system, interdiscipline communications, etc.

Case studies provide a useful format in which to investigate errors. These studies should include instances of success and near failure, in order to identify the important parameters. Case studies may be used to develop theoretical error models, serve as educational tools for colleges, and provide practicing engineers a source of reference material.

Specific steps can immediately be taken to reduce errors. They primarily include the implementation of appropriate and adequate quality control measures, and the use of design checklists and fact sheets. Checklists can further be implemented within computer programs.

Development of error control strategies depends upon accurate modeling of the processes of design, construction and use, and their error occurrences, consequences, and control. Only then can engineers make rational decisions among design and construction alternatives.

# HUMAN ERROR IN DESIGN AND STRUCTURAL FAILURE

Ruediger Rackwitz*

## Introduction

According to Matousek and Schneider (1975) and others, roughly 9 of 10 serious structural accidents can be traced back to human error in some form. Again, roughly half of those errors occur during planning and design. It, therefore, makes sense to study the problem of human error in this phase of the building process. Unfortunately, the data basis for the realistic modelling of occurrences, type and size or effect of errors and for their detection during some control action is extremely poor. But it will remain so unless some reasonable yet crude models indicate at least the type of data to be collected and, further, how to statistically evaluate such data in order to quantify the parameters of the model. Many aspects of human error generations and detection are, in fact, known in qualitative terms and are touched upon in many relevant studies (see, for example, the material collected in IABSE (1983), IABSE (1985), IABSE (1986) and in the proceedings of this workshop) but only very few studies address the problem of converting the qualitative understanding into numerical models (see, however, the contributions work by Melchers and Nowak and Lind in IABSE (1985)).

In the following we shall study in a rather speculative and partially simplistic manner the dependence structure of error occurrence, error detection and structural failure. Furthermore, some useful analysis techniques will be discussed.

## Structural Failure and Error Model

Human errors can affect both structural resistance and the actions resp. the action effects of a structure. On a component level one may symbolically write the failure event as:

$$V = \{E\ R - F\ \ S \leq 0\} \tag{1}$$

where R is the capacity, S the load effect, E a factor modelling the error affecting the capacity and, similarly, F a

*Technical University of Munich, Munich, Federal Republic of Germany

factor modelling the error affecting the load effect. For simplicity, we shall not distinguish between errors affecting the capacity or the demand. Then, a simpler formation is:

$$V = \{E \frac{R}{S} < 1\} = \{\ln E + \ln R - \ln S \leq 0\} \qquad (2)$$

Several errors can occur in a design and it is necessary to consider their interaction. A simple assumption is that they combine in a multiplicative manner, that is

$$E = \prod_{j=1}^{n} E_j \qquad (3)$$

with which the failure event reads:

$$V = \{\prod_{j=1}^{n} E_j \, R/S \leq 1\} = \{\sum_{j=1}^{n} \ln E_j + \ln R - \ln S \leq 0\} \qquad (4)$$

The values of the $E_j$'s are unity in an error free situation. Errors usually have no direction. Also it is assumed that we have $0 \leq E_j < E_{ej}$ with $E_e$ a reasonable upper limit. There can be many arguments about the type of distribution function of the $E_j$'s. These must depend on the type of error but also on the psychology of its generation. There are reasons to believe that the distribution of the effect, $E_j$, of an error is rather widespread although the mental deviation from the correct action during its generation may not be large. In absence of more detailed information we simply take a uniform distribution. In order to take account of the argument that errors are somewhat more likely to decrease structural safety than to increase it, we set the above mentioned upper limit at $E_e = 1.5$. This implies that on the average one error reduces the safety ratio R/S by 25%. It also implies that two or more errors use up the nominal safety ratio aimed at in traditional, deterministic design.

Further, it appears reasonable to distinguish between two types of errors although there cannot always be a clear-cut differentiation. Type 1 comprises design mistakes due to inadvertence such as simple arithmetic errors, errors in the input or output data, misreadings and the like. Type 2 errors are conceptual errors, i.e., the omission of an important load or hazard scenario, the misinterpretation of soils data, inadequate structural lay-out and/or inadequate modelling of the load-bearing structural system. They especially comprise design errors in tasks which are unfamiliar to the specific designer but otherwise are standard from a professional point of view. Type 1 usually is easy to detect because the data and operational environment is given and known. Formal tools are available to detect those errors, e.g., rereading of input or output data, equilibrium and compatibility checks. Type 2 is much more difficult to detect. But it occurs less frequently. For simplicity, it is assumed that the effects of both types of errors are the same.

Error Occurrence and Error Detection

The occurrence of errors should be a rare event in the sequence of intellectual operations carried out during design. In accordance with most of the literature on this subject, a Poisson process is assumed which implies that the mean error content in a design increases linearly with its size or, more precisely the number of error-prone intellectual operations.

Let, then, $m = \mu s = (\mu_1 p_1 + \mu_2 p_2)s$ be the total expected number of errors and $p_1$ and $p_2$ the proportions of type 1 and type 2 errors, respectively $(p_1 + p_2 = 1)$, and s denotes the size of a project. The probability distribution of k errors in a project is, of course the Poisson distribution. The probability of having at least one error is:

$$P(N_E \geq 1) = 1 - \exp[-\mu s] \tag{5}$$

For the detection process, we assume a non-homogeneous Poisson process with mean value function (see, for details, Yamada and Osaki 1984)

$$m_i(t) = p_i m(1 - e^{-b_i t}) \tag{6}$$

with $m_i(o) = 0$, $m_i(\infty) = p_i m$ and t is the checking time. Following the description of the two types of errors we assume

$$0 \leq b_2 \leq b_1 \leq 1$$

The error detection process has intensity function

$$\lambda(t) = \frac{dm(t)}{dt} = m \sum_{i=1}^{2} p_i b_i e^{-b_i t} \tag{7}$$

with $m(t) = m_1(t) + m_2(t)$. It also follows that the detection rate per error is

$$d(t) = \frac{\lambda(t)}{m - m(t)} =$$

$$= \sum_{i=1}^{2} \left(\frac{p_i e^{-b_i t}}{p_1 e^{-b_1 t} + p_2 e^{-b_2 t}}\right) b_i \tag{8}$$

which is a decreasing function of time and $d(o) = p_1 b_1 +$

$p_2b_2$ and $d(\infty) = b_2$. As expected, in the late phase of checking there are primarily type 2 errors still to be eliminated and this is done with almost constant rate close to $b_2$. The number of remaining errors at checking time, t, can also be shown to be Poisson distributed with mean $\kappa = m - m(t)$.

Hence, the probability of having exactly k errors remaining in the design is given by:

$$P(\underline{N}_E = k) = \frac{\kappa^k}{k!} \exp[-\kappa] \qquad (9)$$

Note that $m(\infty) = p_1m + p_2m$ and, therefore, the checking time is infinite in order to remove all errors with probability one.

Parameters For the Error Occurrence Detection Model
Error Occurrence

The error occurrence parameters $\mu_1$ and $\mu_2$ depend on many influences only a few of which will tentatively be modelled herein. First of all, the parameters vary between designers and, probably, even for the same designer from project to project. These variations reflect the various intellectual capabilities, the psychological dispositions, the working environment, the involvement in the profession and a number of other factors. These variations may be taken into account by a random variable Q whose average probably is in the order of $10^{-3}$ to $10^{-4}$ [errors per ]Aoperation] but with large coefficient of variation in the order of 50%, say. The error rate increases with time pressure. Furthermore, the error may vary with the professional age A, i.e., the professional experience of the designer but, presumably, in a different way with respect to type 1 and type 2 errors. As a first attempt, let us use:

$$\mu_1 = Q\, b_1^{A_1}\, \bar{\tau}/\tau \qquad (10)$$

is a constant in the order of 0.98 implying that the error rate at professional age $A_1 = 40$ (years) goes down to 45% of the error rate of a beginner, $\bar{\tau}$ is the time a designer would need to perform his tasks with usual care. $\tau$ is the time he is actually allowed to use. For a specific project one has to assume that the professional age of the designer is a random variable. With reference to our definition of type 1 errors a triangular distribution with density $f_1(a) = 2\, a/a_m^2$ and $a_m$ set at 40 is assumed. This distribution reflects the fact that young engineers are more likely to be involved in standard design tasks.

For type 2 errors we use instead

$$\mu_2 = Q \; b_2^{A_2 \bar{\tau}/\tau} \; c^m \tag{11}$$

which is similar to eq. (10). But, as individual experience grows, the probability of conceptual errors decays more rapidly with professional age. Therefore, the constant $b_2$ may tentatively be set at 0.96 implying that at $A_2 = 40$ a designer has reduced his error rate down to 20%. We expect that senior engineers are more likely to be involved in the type of conceptual decisions we are concerned with. Again, a triangular distribution is assumed but now with the largest density at $a = a_m$. It can be written as: $f_2(a) = 2(a_m - a)/a_m^2$. There is general agreement that conceptual errors can be avoided by post-graduate education and training. The last factor shall take account of this tendency. Herein, c is a constant around 0.96 and m is the number of courses attended, say $m = A_2/2$. This means that experienced engineers at the end of their active professional life have an error rate of less than 10% of a beginner's error rate which appears reasonable.

Error Detection

The primary detection parameters are the parameters $b_1$ and $b_2$ and the checking time t. For $b_i$ (i = 1,2) we make the "Ansatz":

$$b_i = P_i \left(1/\prod_{j=1}^{n} E_j\right)^{\nu_i} \tag{12}$$

$$\nu_i = \begin{cases} 0{,}2 & \text{for } i = 1 \\ 0 & \text{for } i = 2 \end{cases}$$

The last factor takes account of the fact that type 1 errors appear to be more easily detected if they are large. The product term reflects the tendency of the controlling engineer to be more careful once one unfavorable error has been detected. The exponent $\nu_2$ has been set equal to zero for conceptual errors because the existence of those errors is not evident from a check of structural dimensions and the like. Similar to the occurrence rates for type 1 and type 2 errors we assume that there is on the average $P_1/P_2 = 10$. Furthermore, since checking is normally done by experienced engineers, we assume $P_1 = \mu_1 Q$ = (10 to 100) Q resp. $P_2 = \mu_2 Q$ = (1 to 10) Q and $t \approx s/3$.

However, design and checking engineers tend to rely on each other in the sense that the latter checks more carefully if he knows that the designer is below average and vice versa. Hence, the detection parameter should also depend on the actual error parameter.

This may be modelled as follows. Let Q be log-normally distributed and represented by:

$$Q = m_Q \exp [U_Q \sigma_Q] \tag{13}$$

Then, the random log-normal variables $P_1$ and $P_2$ can be represented by

$$P_1 = m_Q \mu_1 \exp [(\rho_1 U_Q + \sqrt{1 - \rho_1^2}\, U_{P_1}) \sigma_1] \tag{14a}$$

$$P_2 = m_Q \mu_2 \exp [(\rho_2 U_Q + \sqrt{1 - \rho_2^2}\, U_{P_2}) \sigma_2] \tag{14b}$$

with $\sigma_1$, $\sigma_2$ the logarithmic standard deviation and $\rho_1$, $\rho_2$ the correlation coefficients. Values of $\rho_1 \approx \rho_2 \approx 0.5$ appear reasonable. A larger correlation coefficient for type 2 errors might model the observation that the dependence is stronger for conceptual design decisions (design and checking engineers educated at the same school of thought).

System Analysis

The formal structure of the system can now be given. System failure is defined as:

$$\begin{aligned} F &= [\{N_E = 0\} \cap \{V(N_E = 0)\}] \\ &= [\{N_E \geq 1\} \cap \{\bar{N}_E = 1\} \cap \{V(\bar{N}_E = 1)\}] \\ &= [\{N_E \geq 1\} \cap \{\bar{N}_E = 2\} \cap \{V(\bar{N}_E = 2)\}] \\ &= [\{N_E \geq 1\} \cap \{\bar{N}_E = 3\} \cap \{V(\bar{N}_E = 3)\}] \\ &= \ldots \end{aligned} \tag{15}$$

Unfortunately, the intersection probabilities are not easy to determine since the events in the system are dependent due to common uncertain parameters. An exception is the first line in Eq. (15) representing the error-free case. Here the events $(N_E = 0)$ and $V(N_E = 0)$ are independent. A numerical solution is possible with modern first- or second-order reliability methods. These require the formulation of the events in the form $g_i(\underline{U}) \leq 0$ with $\underline{U}$ the vector of transformed uncertain variables. This can readily be achieved by introducing auxiliary variables, i.e., by setting

$$P(N_E \geq 1) \equiv 1-\exp [-\mu s] = 1- \Phi(U_E) = P\,(U_E \geq U_E) \tag{16}$$

we have:

$$\begin{aligned} \{N_E \geq 1\} &= \{-U_E - \Phi^{-1} [1-\exp [-\mu s]\} \leq 0\} \\ &= \{- \ln \Phi\, (-U_E) - \mu s \leq 0\} \end{aligned} \tag{17}$$

The transformation of the event $\{N_E = k\}$ is more difficult. We write

$$\{\bar{N}_E = k\} = \{\bar{N}_E \leq k\} \cap \{\bar{N}_E > k - 1\} \tag{18}$$

and, therefore, apply the identities:

$$P(\bar{N}_E \leq k) = \Phi(U_{\bar{E}_1}) \tag{19a}$$

$$P(\bar{N}_E > k - 1) = 1 - \Phi(U_{\bar{E}_2}) \tag{19b}$$

With these manipulations together with some additional numerical provisions, the calculation of the probabilities in Eq. (15) is straightforward.

## Some Numerical Results and Their Interpretation

The models presented before are presently being investigated to some extent. In a separate paper a detailed account of the results will be given. Here only some tentative conclusions from the results available so far will be drawn.

If the parameters are chosen such that the ideal error-free situation corresponds to a failure probability of $4.8\ 10^{-6}$ and that the probability of generating at least one error in the design is approximately 0.5 while the probability of having at least one error remaining in the system after checking is approximately $1.5\ 10^{-4}$, the total failure probability with no error in the system is $2.4\ 10^{-6}$. The total failure probability with precisely one error in the system is $8\ 10^{-6}$. Probabilities with two or more errors in the system are by orders of magnitude smaller and therefore, can be neglected. The error remaining in the system is most likely a type 1 error which appears to be a consequence of the assumption that its generation becomes less likely if senior engineers are involved. The typical professional age (design point age) of that engineer is, in fact, thirty years. If, however, the effect of experience and continued education is less pronounced than as assumed in the section entitled, Error Detection, the remaining errors are primarily of type 2. A study of the equivalent sensitivity values defined as $\partial\beta_{system}/\partial\mu_i$ according to Gollwitzer and Rackwitz (1983) shows that by far the most important parameters are the detection parameter $b_1$ followed by the error size $E_1$. Interestingly, the value of the correlation coefficient between the rates of error generation and of error detection has only a minor influence. On the other hand, the correlation coefficients between the (linearised) failure surfaces in the system obtain non-negligible values. A typical correlation matrix is:

| | $N = 1$ | $N_E < 1$ | $N_E > 0$ | $V(1)$ |
|---|---|---|---|---|
| $N = 1$ | 1 | | symm. | |
| $N_E \leq 1$ | 0.01 | 1 | | |
| $\bar{N}_E > 0$ | -0.01 | -0.2 | 1 | |
| $V\ (1)$ | 0 | 0.28 | -0.28 | 1 |

It explains why in this example the total system failure probability is significantly larger than the product of the individual event probabilities.

In summarizing the results obtained so far one can conclude that focus of further studies should be on error detection models and its parameters. Further, stochastic dependencies among error generation, error detection and structural failure need to be considered in reliability studies. Finally, it is important to model type and size of errors as close to reality as possible.

Summary and Conclusions

The problem of modelling human error occurrence and detection is one of the most important subjects in reliability engineering, probably just inferior to challenge to avoid or at least diminish human errors. At present, there is almost no data. Although any model then becomes debatable, a bad but improvable model is better than none. However, some more involved reliability analyses including human error may not answer the question whether a structure is safe enough or not. It is rather the sensitivity of structural reliability with respect to the various parameters in the entire quality assurance system (QAS) determined by these analyses which offers a means to improve the overall system by distributing the always limited amount of efforts for quality assurance in the most efficient way. Let q be the vector of parameters of a QAS and $P_f(\underline{q})$ the failure probability of the structure. Then, the gradient grad $p_f(\underline{q})$, or with $\beta = -\Phi^{-1}(P_f)$, grad $\beta(\underline{q})$ is such a sensitivity measure which contains essential information on how to modify the system. In fact, as pointed out several times, the design of efficient QAS's is an optimization task with overall reliability as the objective and total effort for quality and, possibly, some minimum reliability as constraints. There may be QAS's which are better than others in this respect.

References

Gollwitzer, S. and Rackwitz, R., "Equivalent Components in First-Order System Reliability", Reliability Engineering, Essex, Vol. 5, 1983.

IABSE, Workshop on Quality Assurance within the Building Process, Rigi, 1983, IASBE-Report, Vol. 47, 1983.

IASBE, Symposium on Safety and Quality Assurance of Civil Engineering Structures, Tokyo, 1986, IABSE-Report, Vol. 50, 1985.

IABSE, Symposium on Safety and Quality Assurance of Civil Engineering Structures, Tokyo, 1986, IABSE-Report, Vol. 51, 1986.

Matousek, M and Schneider, J., Untersuchungen zur Struktur des Sicherheitsproblems bei Bauwerken, Bericht Nr. 59, Institute fur Baustatik and Konstruktion, Zurich, 1975.

Yamada, S and Osaki, S., "Nonhomogeneous Error Detection Rate Models for Software Reliability Growth", Proc. Symp. on Stochastic Models in Reliability Theory, Nagoya, 1984, in: Lecture Notes in Economics and Mathematical Systems, Springer, Berlin, 1984, pp. 120-143.

## List of Participants

| | |
|---|---|
| David E. Allen | National Research Council<br>Ottawa, Ontario, Canada |
| Gita Amiri | University of Maryland<br>College Park, MD |
| Abdulrahim M. Arafah | University of Michigan<br>Ann Arbor, MI |
| David I. Blockley | University of Bristol<br>Bristol, England |
| Colin B. Brown | University of Washington<br>Seattle, WA |
| Robert I. Carr | University of Michigan<br>Ann Arbor, MI |
| Neal FitzSimons | Engineering Counsel<br>Kensington, MD |
| Dan M. Frangopol | University of Colorado<br>Boulder, CO |
| Ted V. Galambos | University of Minnesota<br>Minneapolis, MN |
| Mircea D. Grigoriu | Cornell University<br>Ithaca, NY |
| Fabian C. Hadipriono | Ohio State University<br>Columbus, Ohio |
| W. Brent Hall | University of Illinois<br>Urbana, IL |
| Robert D. Hanson | University of Michigan<br>Ann Arbor, MI |
| Owen G. Ingles | Consultant<br>Tasmania, Australia |
| Franz Knoll | Nicolet-Chartrand-Knoll<br>Montreal, Quebec, Canada |
| Niels C. Lind | University of Waterloo<br>Waterloo, Ontario, Canada |
| Robert E. Melchers | University of Newcastle<br>New South Wales, Australia |

| | |
|---|---|
| Juan A. Melgarejo | University of Michigan<br>Ann Arbor, MI |
| Janusz W. Murzewski | Politechnika Krakowska<br>Krakow, Poland |
| Andrzej S. Nowak | University of Michigan<br>Ann Arbor, MI |
| Nick F. Pidgeon | University of Bristol<br>Bristol, England |
| Ruediger Rackwitz | Technische Universitaet<br>Munich, West Germany |
| M.K. Ravindra | EQE Incorporated<br>Newport Beach, CA |
| Heki Shibata | University of Tokyo<br>Tokyo, Japan |
| Norris Stubbs | Texas A & M University<br>College Station, TX |
| Carl J. Turkstra | Polytechnic Institute of New York<br>Brooklyn, NY |
| Y-K. Wen | University of Illinois<br>Urbana, IL |

## SUBJECT INDEX

Page number refers to first page of paper.

## AUTHOR INDEX

Page number refers to first page of paper.